AF877057

Crystal Research and Technology

12

Journal of Experimental and Industrial Crystallography

Zeitschrift für experimentelle und technische Kristallographie

Volume 22
1987
Number 12 December

Akademie-Verlag · Berlin

ISSN 0232-1300 Cryst. Res. Technol., Berlin 22 (1987) 12, 1443–1580, K219–K240

Directions to the author

In the journal "Crystal Research and Technology – Kristall und Technik" as a rule the following categories of papers will be printed: original papers, review articles, short notes, and book reviews, preferentially in **English**. Complete information necessary for type-setting and a "Short Title" should be added.

The manuscript should not exceed 10 pages and be type-written on one side only in the format A4 (21 × 29.7 cm) on white ink-proof paper, with a two-lined distance and a left-hand free edge of 3 cm, and submitted in 2 copies. All sheets of the paper belonging to the contribution are continuously to be numbered.

The head of the manuscript should be arranged in this manner: initials and name of the author; institution, and title of the paper. Then two abstracts should follow, one in the original language and the second in German, English, French, or Russian, respectively.

Formulae, signs and symbols, letters etc. being wanting on the type-writer, should be distinctly written with penholder. O as zero is to be underlined, the capital O is not. Greek letters should be underlined with red ink.

Paragraphs should be marked by indenting of at least 3 touches.

Concerning the marking of the manuscript should be noted:
headings should be underlined, sub-headings not, and indented by 9 touches. The titles of tables are not to be underlined. Other markings in the text will be done by the editorial staff. They mean:

light-red 〰〰〰 = small capitals
blue 〰〰〰 = in italics

light-red | perpendicular straight red line on the left edge = small printing

Tables are to be added on separate sheets of paper at the end of the contribution.

Foot notes should be continuously written on a separate sheet of paper.

Figures (Fig.) – drawings and photographs – should be not more than 8 and submitted (2 copies) in a special envelope: photographs as white high-gloss prints (inscription on a transparent covering sheet) and drawings as distinct sketches. On the back of the submitted figures the reduction scale in tenth (9/10, 8/10, 5/10 e.g.), author's name, short title of the paper, and the number of the figure should be given.

Captions for the figures should be submitted on a separate sheet of paper.

References should be written without numbers in alphabetical order of the names at the end of the contribution. Cited names in the text stand for the references. The references should contain the following items, for example:

for books: RAST, K.: Physikalisch-chemische Rechnungen, Berlin 1958
for journals: KRAUSE, H. N., MÜLLER, K.: Cryst. Res. Technol. **16** (1981), 70

Add, please, the short title, about 6 keywords, and the academic grades of the authors.
The authors will be sent two galley proofs of review articles and original papers. Proof reading of short notes and book reviews will be done by the editorial staff.

For each of these papers a total of 50 reprints is available free of charge. Requests for additional maximum 200 reprints (against payment) should be given together with the submitted manuscript.

12

Crystal Research and Technology

Founded 1966 as
Kristall und Technik

Journal of Experimental and Industrial Crystallography

ESTABLISHED BY
W. KLEBER and H. NEELS

EDITED BY
H. NEELS, Leipzig

EDITOR-IN-CHIEF
H. NEELS, Leipzig

ADVISORY BOARD
L. N. ALEKSANDROV, Novosibirsk
Č. BARTA, Prague; F. BERTAUT, Grenoble
G. BLIZNAKOV, Sofia; J. BOHM, Berlin
O. BRÜMMER, Halle
V. A. FRANK-KAMENETZKIJ, Leningrad
S. HAUSSÜHL, Köln
J. HEYDENREICH, Halle
R. KAISCHEW, Sofia; R. KERN, Marseille
Y. KOMURA, Hiroshima; I. KOSTOV, Sofia
P. KRISHNA, Varanasi
K. ŁUKASZEWICZ, Wrocław; K. MEYER, Berlin
P. PAUFLER, Leipzig; I. S. REZ, Moscow
W. SCHATT, Dresden, N.N. ŠEFTAL', Moscow
D. SCHULZE, Dresden; N.N. SIROTA, Moscow
I. TARJAN, Budapest; O.M. TODES, Leningrad
A. A. TSCHERNOV, Moscow
A. F. WITT, Cambridge (U.S.A.)
W. A. WOOSTER, Cambridge

VOLUME 22 · 1987 · NUMBER 12

Akademie-Verlag · Berlin

Papers should be directed to the editor's office
DDR-7030 Leipzig, Scharnhorststraße 20,
or to the editor Prof. Dr. Hermann Neels,
DDR-7254 Machern, Zeititzer Weg 25,
or to one of the members of the Advisory Board.

Orders can be sent

- in the GDR:
  to the Postzeitungsvertrieb or to the Akademie-Verlag Berlin, Leipziger Straße 3–4, DDR-1086 Berlin;
- in the other socialist countries: to a book-shop for foreign language literature or to the competent news-distributing agency;
- in the FRG and Berlin (West): to a book-shop or to the wholesale distributing agency Kunst und Wissen, Erich Bieber OHG, Wilhelmstr. 4–6, D-7000 Stuttgart 1;
- in the other Western European countries: to Kunst und Wissen, Erich Bieber GmbH, General Wille-Str. 4, CH-8002 Zürich;
- in USA and Canada to the VCH Publishers, Inc., 303 NW 12th Avenue, Deerfield Beach, FL 33442-1788, USA;
- in other countries. to the international book- and journal-selling trade, to Buchexport, Volkseigener Außenhandelsbetrieb der DDR, Postfach 160, DDR-7010 Leipzig, or to the Akademie-Verlag Berlin, Leipziger Straße 3–4, DDR-1086 Berlin.

Journal „Crystal Research and Technology"

Herausgeber: Im Auftrage des Akademie-Verlages Berlin herausgegeben von Prof. Dr. Hermann Neels.

Verlag: Akademie-Verlag Berlin, Leipziger Str. 3–4, DDR-1086 Berlin; Fernruf: 2236221 oder 2236229; Telex-Nr.: 114420; Bank: Staatsbank der DDR, Berlin, Kto.-Nr. 6836-26-20712.

Chefredakteur: Prof. Dr. Hermann Neels, Zeititzer Weg 25, DDR-7254 Machern; Fernruf: Brandis 3442. Anschrift der Redaktion: Scharnhorststr. 20, DDR-7030 Leipzig; Telefon: 310502.

Veröffentlicht unter der Lizenznummer 1526 des Presseamtes beim Vorsitzenden des Ministerrates der Deutschen Demokratischen Republik.

Gesamtherstellung: VEB Druckerei „Thomas Müntzer", DDR-5820 Bad Langensalza.

Erscheinungsweise: Das Journal „Crystal Research and Technology" erscheint monatlich. Die 12 Hefte eines Jahrganges bilden einen Band. Bezugspreis je Band 480,– DM zuzüglich Versandspesen; Preis je Heft 40,– DM. Der gültige Jahresbezugspreis für die DDR ist der Postzeitungsliste zu entnehmen.

Bestellnummer dieses Heftes: 1082/22/12.

Urheberrechte: Alle Rechte vorbehalten, insbesondere die der Übersetzung. Kein Teil dieser Zeitschrift darf in irgendeiner Form – durch Photokopie, Mikrofilm oder irgend ein anderes Verfahren – ohne schriftliche Genehmigung des Verlages reproduziert oder in eine von Maschinen, insbesondere von Datenverarbeitungsanlagen verwendbare Sprache übertragen oder übersetzt werden. All rights reserved (including those of translations into foreign languages). No part of this issue may be reproduced in any form, by photoprint, microfilm, or any other means, nor transmitted or translated into a machine language, without written permission from the publishers.
© 1987 by Akademie-Verlag Berlin. Printed in the German Democratic Republic.
AN (EDV) 35830
02800

Cryst. Res. Technol.	22	1987	12	1445–1451

Original Papers

H. Kühne, W. Malze

Akademie der Wissenschaften der DDR, Institut für Halbleiterphysik, Frankfurt (Oder), and VEB Halbleiterwerk Frankfurt/Oder

On the Deposition Mechanism of Boron in Silicon CVD Epitaxy

The deposition mechanism of boron doping in CVD silicon epitaxy has been investigated by exposing silicon substrates to B_2H_6–H_2 doping gas mixtures at epitaxy temperatures and examining the effect by dopant profile measuring in an afterwards intrinsically in-situ deposited epitaxial silicon layer. It has been shown that boron is deposited increasing its concentration on the surface linearly with prolonged exposition time and desorbed by purging the surface in pure hydrogen. In the latter case its content decreases linearly proportional to the predeposited concentration. The desorbed boron builds up a secondary doping source which maintains a parasitic boron flow for reincorporation during following layer growth.

Es wurde der Mechanismus des Boreinbaus während der Abscheidung von epitaktischen CVD Siliciumschichten untersucht, indem Siliciumsubstrate der Einwirkung von B_2H_6–H_2 Dotiergasmischungen bei Epitaxietemperaturen ausgesetzt wurden. Der erreichte Effekt wurde durch die Auswertung der unmittelbar anschließend in einer dünnen und intrinsisch abgeschiedenen Siliciumschicht erhaltenen Dotierungsprofile gemessen. Danach nimmt die Oberflächenkonzentration von Bor in der Vorabscheidungsphase linear mit der Expositionszeit zu und während einer Spültemperung in reinem Wasserstoff linear proportional zur Vorabscheidungskonzentration ab. Das desorbierte Bor gibt Anlaß zur Bildung einer Sekundärquelle die während einer nachfolgenden Schichtabscheidung einen parasitären Boreinbau verursacht.

Introduction

Intentional doping and autodoping of epitaxial silicon during layer growth has been interpreted in general but also especially with respect to boron doping by suggesting an adsorption-type mechanism (Ducheman; Pogge et al.; Tabe, Nakamura; Reif et al.; Srinivasan; Richter, Kühne). Contrary to arsenic and phosphorus both modes of boron doping, intented and unintented doping, show some pecularities characterized, for instance, by a positive sign of temperature dependence, strong dependence on growth rate at high deposition temperatures (Wagin, Skworzow), and lack of autodoping suppression under low pressure deposition conditions (Kulkarni, Kozul). By those experimental findings less an adsorption-like incorporation behaviour than an active deposition behaviour is announced rather.

In the present paper we make an attempt to bring matters nearer to a definite decision.

Theoretical background

Adsorption of dopants should not necessarily be linked up with silicon layer growth even not when the dopant source must be transformed into element atoms by means of chemical reactions. Though it should be characterized by a time-dependent growth of adsorbate concentration when dopants are offered from the gas at a constant level, and, in the contrary, it should be characterized by a decrease of dopant adsorbate concentration when the loaded silicon surface is heated in a dopant-free gas stream.

The idea is that a certain dose of dopants being adsorbed on the silicon substrate surface will built up a small dopant profile in an epitaxial silicon layer intrinsically grown on the predeposited silicon surface. Such a profile could be measured by applying known CV-methods provided that its maximum concentration does not exceed the limit of about 1 E 16 atoms · cm^{-3}. In case of solid state diffusion a gaussian-distributed profile will be expected when a remarkable but finite dose on the substrate surface acts as its origin. It should be characterized by an erfc-distribution when the constant background doping of the substrate has to be considered as the source. Yet, we would find an exponential falling off of the doping level when an active redistribution takes place during layer growth. Taking into account a constant background dopant level of the deposition system the profile can be described by the following equation:

$$[N(x) - N_\infty] = [N_0 - N_\infty] \cdot \exp - \frac{\Delta x}{L}, \tag{1}$$

where

$N(x) - N_\infty$	doping profile
N_∞	background doping level
N_0	maximum doping level at the substrate-layer boundary
Δx	distance from the substrate surface into the silicon layer
L	characteristic length.

That part of the originally on the surface deposited boron dose which has been redistributed in the epitaxial layer may be estimated from equ. (1) by integration within the limits $N(x) = N_0$ and $N(x) = N_\infty$. That mathematical manipulation can be carried out in a simple way when the limits are approached by $\Delta x = 0$ and $\Delta x = \infty$. The result is given by the following expression:

$$Q_{B,\text{epit}} = L \cdot (N_0 - N_\infty) \,. \tag{2}$$

That other part of boron deposit that has been moving into the substrate by means of solid state diffusion forming a counter part of the profile, should be obtained by integrating the gaussian function taking the constant doping level of the substrate into account.

$$[N(x) - N_{\text{sub}}] = [N_0 - N_{\text{sub}}] \cdot \exp - \frac{\Delta x^2}{4 \cdot D \cdot t} \tag{3}$$

where

N_{sub}	doping level of the substrate
Δx	distance from the substrate surface into the substrate.

The distributed boron dose then is as follows:

$$Q_{B,\text{sub}} = \sqrt{\pi \cdot D \cdot t} \cdot [N_0 - N_{\text{sub}}] \,. \tag{4}$$

Experiments

Boron doped wafers (orientation 100, $N_{\text{sub}} \approx 9$ E 14 cm^{-3}) were heated in a hydrogen atmosphere toward about 1325 K and exposed to a diborane-hydrogen gas mixture streaming for different periods of time through the reaction chamber (c.p. Table 1). In one of the experimental runs silicon layer growth was started immediately after the diborane vent had been closed. In another run layer deposition began not till the end of a three minutes purging with pure hydrogen. Layers 1.2 µm in thickness were intrinsically deposited from a silane-hydrogen-hydrogen chloride mixture characterized by a layer growth rate of about 0.3 µm · min^{-1}. Layer deposition was followed by a thermal oxidation step. A 30 nm SiO_2 layer was grown at 1173 K in an oxygen-hydrogen chloride atmosphere. Interface charges and fast states were annealed by a treatment combination in nitrogen at 1273 K for 30 minutes and in hydrogen at 673 K for 60 minutes.

The wafers prepared in the way described above underwent a CV-measuring procedure by applying a mercury-electrode system in connection with a capacity measuring bridge (measuring frequency: 1 kHz, measuring range: 100 pF to 100 nF) and a voltage rampe generator (range: −100 V to 100 V, rampe: 2 mV · s^{-1} to 5 V · s^{-1}).

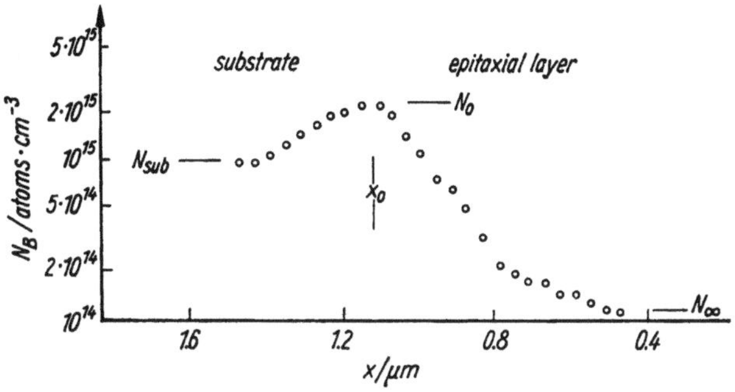

Fig. 1. Boron profile evaluated from CV-measurements, B_2H_6–H_2 exposure time: 12 minutes, no H_2-purge

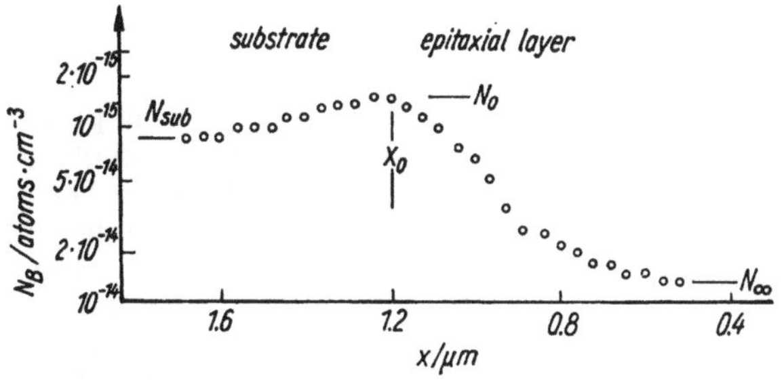

Fig. 2. Boron profile evaluated from CV-measurements, B_2H_6–H_2 exposure time: 12 minutes, H_2-purge: 3 minutes

The Figures 1 and 2 show typical examples of the profile plots obtained. At the substrate-layer boundary (x_0) a boron concentration peak (N_0) is clearly visible rising above the substrate doping level. The boron profile decays towards a background doping level of $N_\infty = (1-2)$ E 14 cm^{-3}. Though similar in general both profiles are different in the boron peak concentration on the one hand and in the steepness of the falling off on the other. Those differences are caused by applying a cleaning purge or not.

The profile measuring points were evaluated by means of equ. (1). Its logarithmated version allows a straight-line representation of each of the measured profiles (Fig. 3) from which the characteristic length L can be obtained by forming a simple difference quotient $[\Delta x/\Delta \ln (N - N_\infty)]$. The term L is necessary for calculating the boron dose distributed in the epitaxial layer (c.p. equ. (2)) and, in addition, allows the estimation of the decay time constant τ of the system considered (REIF et al.) when the layer growth rate is known:

$$\tau = \frac{L}{v_{epi}} \text{(min)} . \quad (5)$$

Table 1 contains values of both quantities L and τ as well as such for the difference of the peak and background concentrations, the epitaxially distributed boron doses, and the position of the substrate-layer boundary (x_0) measured from the layer surface.

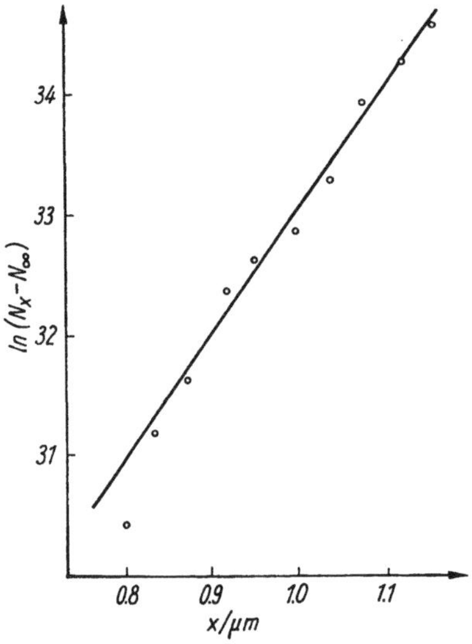

Fig. 3. Semilog-plot of boron profile; B_2H_6-H_2 exposure time: 6 minutes, no H_2-purge. x_0 = 1.2 µm, $(N_0 - N_\infty)$ = = 1.8 E15, L = 0.096 µm

Discussion

The characteristic length L will increase in such an extent as the profile decays less steep from the peak concentration to the background doping level of the layer. With a cleaning purge inserted in the preparation sequence between the boron predeposition step and the layer growth step, the amount of L is doubled in comparison with preparations missing that purge (c.p. Table 1). This result has been totally unexpected, though it is a reasonable finding. Recently it has been concluded by ARNOLD from a critical discussion of the amounts of the decay time constants measured in various deposition systems with respect to those expected for different reaction mechanisms that not only substrate wafers but also substrate holders and reaction chamber walls may likewise act as sources of parasitic doping phenomena showing non-steady state behaviour.

So, in connection with the investigated problem attention should be given to the transfer behaviour of the layer deposition system having been boron predeposited to different degrees and also to the behaviour of unloaded but to different levels doped wafers in a clean deposition system.

The more important result of our investigation is shown in Figure 4. The maximum boron concentration $(N_0 - N_\infty)$ has been selected as a measure of the boron predeposition efficiency. The figures illustrate the fact that the boron content in the layer increases linearly proportional to the exposition time. That is typical for adsorption processes in the range of scarcely covered surfaces. A cleaning purge in hydrogen reduces the originally deposited boron content on the wafer surface, certainly, in its degree, depending on the extent of the purge step. In Figure 4 the 1E15 cm^{-3} level is considered the limit below which the obtained effects are mainly caused by substrate doping.

Figure 5 shows the boron dose found within the epitaxial layer versus the exposition time. The linearity of the purge-free experimental run can clearly be dis-

Table 1

Nr.	t_{B2H6}/min	t_{H2}/min	$x_0/\mu m$	$(N_0 - N_\infty)$/atome · cm^{-3}	$L/\mu m$	τ/min	Q_B/atome · cm^{-2}
1	3	—	1.2	9 E 14	0.119	0.4	1.1 E 10
2	1.5	—	1.3	9.3 E 14	0.099	0.33	0.93 E 10
3	6	—	1.2	18 E 14	0.096	0.32	1.7 E 10
4	12	—	1.1	26.4 E 14	0.127	0.42	3.3 E 10
5	3	3	1.2	9.2 E 14	0.201	0.67	1.8 E 10
6	1.5	3	1.3	7.1 E 14	0.220	0.73	1.6 E 10
7	6	3	1.3	9.2 E 14	0.183	0.61	1.7 E 10
8	12	3	1.2	11.9 E 14	0.213	0.71	2.5 R 10

cerned. A similar behaviour is indicated, but not clearly visible, for the run containing a purge step. The desorption effect of that cleaning purge is not as clear as in Figure 4 because of an additional boron incorporation during layer growth. The latter parasitic doping effect causes an oversize of the boron dose as can be seen in those examples

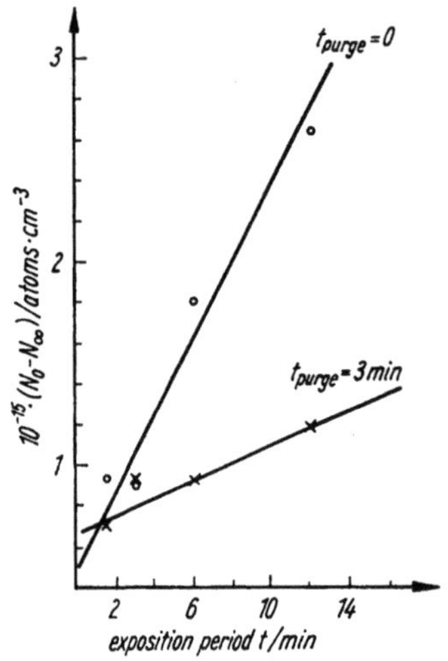

Fig. 4. Boron peak concentration at substrate layer-boundary versus B_2H_6–H_2 exposure time

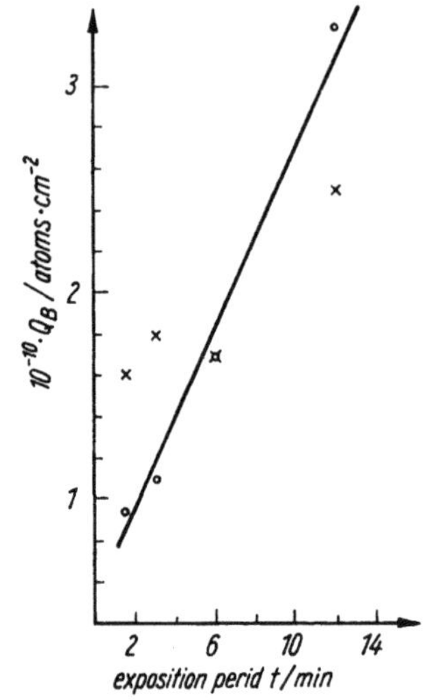

Fig. 5. Boron dose distributed in the epitaxial layer versus B_2H_6–H_2 exposure time. ... without H_2-purge; 3 minute H_2-purge

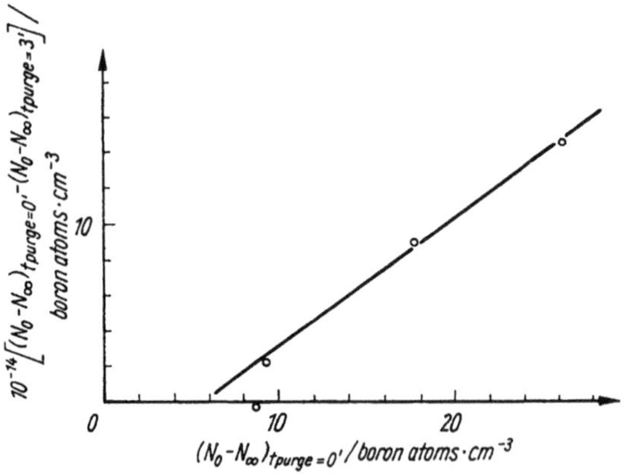

Fig. 6. Boron desorption behaviour illustrated by the difference of boron peak concentrations before and after the application of a H_2-purge step as a function of the original boron peak concentration

where exposition times shorter than six minutes are applied. This oversize of dose has been built in in spite of the decrease of the boron peak concentration at the substrate-layer boundary (c.p. Fig. 4).

Desorption of boron during the purge step proceeds, as should have been expected, proportional to the predeposited boron dose or, more precisely formulated, proportional to the peak concentration of boron being the result of exposing the substrate surface to diborane. Figure 6 demonstrates the described desorption behaviour.

The remarkable extention of the doping profiles into the epitaxial layer (c.p. Figs. 1 and 2) may be a proof of an active dopant redistribution in the initial period of layer growth occurring beside of solid state diffusion and evaporation. The problem considered in the paper needs further investigations. First of all, a comparing examination of arsenic and phosphorus has to be carried out regarding their zero-layer-growth behaviour. On the other hand the influences of the process parameters should be clearly revealed and the predeposition behaviour of boron inspected not only in the range of a scarcely covered surface but also up to monolayer concentrations.

References

ARNOLD, H.: Cryst. Res. Technol. **21**, 1421 (1986), **22**, 481 (1987)

DUCHEMAN, J. P.: Rev. Technique Thomson-CSF **9**, 411 (1977)

KULKARNI, S. B., KOZUL, A. A.: Electrochem. Society Extended Abstr. Vol. 80-2, Hollywood (1980), Abstr. 540, 1351

POGGE, H. B., BOSS, D. W., EBERT, E.: Proc. Int. Conf. CVD, The Electrochem. Society, New York 1970, p. 768

REIF, R., KAMINS, T. J., SARASWAT, K.: J. Electrochem. Society **125**, 1860 (1978)

REIF, R., DUTTON, R. W.: ibid. **128**, 909 (1981)

RICHTER, F., KÜHNE, H.: Silicium-Homoepitaxie in: Halbleiterepitaxie, G. SCHNEIDER, L. ICKERT, Leipzig 1984

SRINIVASAN, G. R.: J. Electrochem. Society **125**, 146 (1978), **127**, 1334 (1980)

TABE, M., NAKAMURA, H.: ibid. **126**, 822 (1979)

WAGIN, W. A., SKWORZOV, J. M.: Elektronnaja Technika (russ.), Ser. 2, wyp. **1**, 58 (1979)

(Received May 15, 1987)

Authors' addresses:

Dr. H. KÜHNE
Akademie der Wissenschaften der DDR, Institut für Halbleiterphysik
Walter-Korsing-Str. 2
DDR-1200 Frankfurt (Oder)

Diplom-Chem. W. MALZE
VEB Halbleiterwerk Frankfurt (Oder)
DDR-1201 Frankfurt (Oder)

Book Review

W. Opechowski

Crystallographic and Metacrystallographic Groups

North-Holland, Amsterdam 1986. Pp XIX + 628. Price Dfl. 365.00. In USA/Canada, the b oo is available from Elsevier Science Publ., New York, N. Y. ISBN 0-444-869557

Das eigentliche Anliegen des Autors sind die Gruppen, für die er die Bezeichnung ‚metakristallographisch' einführt um zu unterstreichen, daß ihre Definition logisch nach der der kristallographischen Gruppen vorgenommen werden muß. Die metakristallographischen umfassen dabei Elemente, die auf Funktionen und nicht auf Punkte (wie die kristallographischen) wirken. Derartige Funktionen regeln z.B. die Farbe, die Spin-Anordnung oder die magnetoelektrische Struktur des Kristalls, produzieren also einen endlichen Satz verschiedener Werte. Rund 200 Seiten widmet der Autor den speziellen Beispielen dazu, wobei er ausgiebig auf seine frühere Arbeit über Magnetstrukturen zurückgreift. Der Grundgedanke ist nicht neu, und der Autor knüpft auch direkt an den Pionierarbeiten von H. Heesch an. Der Versuch einer einheitlichen Beschreibung auf allgemeinem Niveau bleibt aber in Anfängen stecken und zerfließt bald in der Diskussion spezieller Phänomene.

Nun sind diesem zweifellos nicht leicht zugänglichen Teil IV drei andere vorangestellt, in denen der Versuch einer Gesamtdarstellung der notwendigen mathematischen und kristallographischen Voraussetzungen gemacht wird (einschließlich Definitionen und Grundtheoremen der Gruppentheorie sowie der Darstellung affiner, euklidischer kristallographischer Gruppen). Der Stil des Autors bringt es nun aber mit sich, daß die Lektüre dieser Einführung ebensolche Anstrengung vom Leser erfordert wie die des erwähnten Hauptteils. Besonderes Kennzeichen ist die vom Autor selbst eingestandene Pedanterie der Begriffserklärung und die zwangsläufig damit einhergehende ungewohnte Symbolik. Aber vor allem das Fehlen von Abbildungen wird der kristallographisch orientierte Leser besonders vermissen. Der Autor gesteht in diesem Zusammenhang, unter dem Einfluß von Lagrange's 'Mécanique analytique' von 1788 gestanden zu haben.

Nach den metakristallographischen Gruppen beschreibt W. Opechowski den Apparat der Superraumgruppen, der als Frucht der Untersuchung inkommensurable Strukturen zu den jüngsten Errungenschaften der Kristallographie gehört und bisher nur in wenigen Lehrbüchern eine Darstellung findet.

Eine ausführliche Bibliographie (mehr als 400 Quellen) und zahlreiche Fußnoten geben dem Werk einen soliden historischen Bezug. Es zeichnet sich durch manche Eigenwilligkeit seines Autors aus und hält dem Eingeweihten in Sachen Gruppentheorie auch manche Anregung bereit. Die hohen Ansprüche an den Leser werden dafür sorgen, daß überwiegend einschlägige Spezialisten der Physik, Chemie und Kristallographie danach greifen.

P. Paufler

Cryst. Res. Technol.	22	1987	12	1453–1458

P. KUMARESON, S. DEVANARAYANAN

Department of Physics, University of Kerala,
Kariavattom, Trivandrum, India

Growth of Crystals of Calcium Sulphate in Silica Gel in Presence of Additive and their Characterisation

Procedure for the growth of calcium sulphate dihydrate crystals in silica gel, in the presence of barium chloride as additive, is described. Optimum conditions for the growth of good quality crystals are worked out. The crystals are characterised by laser Raman spectroscopy, infrared spectroscopy, thermogravimetric analysis and Atomic absorption spectroscopy. It is shown that crystals in the presence of the additive have a greater degree of perfection than those without additive. Amount of water in the crystals is estimated as 20.97%. Presence of calcium and absence of barium in the crystals are indicated.

1. Introduction

Calcium sulphate dihydrate (gypsum) finds extensive use in waste water treatment, desalination and industrial production of phosphoric acid. In crystal form it finds applications in geothermal and petroleum drilling technology. Crystals of calcium sulphate have been grown by spontaneous precipitation method (CHING, MCCARTNEY) and seeded growth technique (LIU, NANCOLLAS). It has been observed that at a given supersaturation and time-interval fewer crystals of calcium sulphate are formed compared to the other alkaline earth sulphates (SLOVENO, TEZAK). Studies have also been made on the precipitation of calcium sulphate in the presence of additives (MCCARTNEY, ALEXANDER) such as citric acid and borax, and the linear growth rate of the crystals in the presence of carboxylic and phosphoric acid derivatives has also been examined (TARDOS, MAYES). In recent years much attention has been given to the growth of crystals by gel-technique, for, it has the added advantage of obtaining macroscopic crystals with perfection of the highest order (HENISCH). A few workers have grown alkaline earth sulphate crystals in silica gel and studied the magnitude and nature of dislocations (PATEL, BHAT; RAJU) and some have grown gypsum crystals in silica gel and agar using acid additives (VAN ROSMALEN et al.). In the present work calcium sulphate crystals have been grown in silica gel in the presence of barium chloride as additive. The crystals have been characterised by spectral and thermal analysis.

2. Experimental

Crystals were grown in silica gel (HENISCH). A.R. grade sodium metasilicate (specific gravity 1.05) was gelled in a U-tube (20 mm ∅ and 150 mm high) at a pH of 8 with 20% sulphuric acid (by volume) which acted as the inner electrolyte. The gel was left for 24 hours to allow it to set firmly. A solution containing equal volumes of barium chloride (0.1 M) and calcium chloride dihydrate (1 M) of both A.R. grade was prepared, and this solution, which formed the outer electrolyte, was carefully added on the top of the gel in one limb of the U-tube. On the other limb about 10 m.l. of 20% H_2SO_4 was taken. The tube was properly sealed and kept undisturbed at room temperature (30 °C). Needle-shaped crystals started growing after a week and the growth process continued for three more weeks. At the solution-gel interface there was a column of white precipitate of length ~ 20 mm and below that thickly populated thin and shining crystals (~ 0.5 mm Ø and 15 mm length) were observed. Though there was sufficient space in the gel for the growth of the crystals, the population density of the crystals was found — as expected — to decrease with the increase of diffusion depth.

Attempts were also made to grow the crystals (i) at high and low *p*H values of the gel, i.e., *p*H = 9 and 7, (ii) at different concentrations of the outer electrolyte and (iii) at different densities of the gel. In all these cases the population density of the crystals was found to drop considerably.

Using the above technique calcium sulphate crystals were also grown in the absence of the additive.

3. Observations and results

3.1. Optical microscopic studies

The crystal was carefully removed and observed through a powerful optical microscope. The crystals grown in the presence of the additive were remarkably long, thin, shining and transparent (Fig. 1) whereas those obtained by growth in the absence of the additive were short, thick, slightly shining and translucent (Fig. 2).

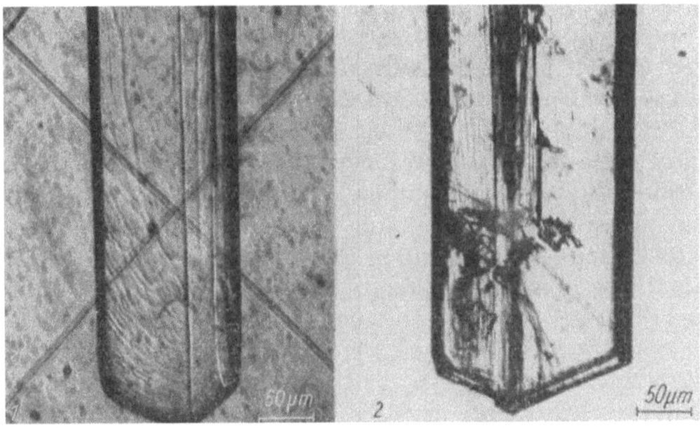

Fig. 1. $CaSO_4 \cdot 2\,H_2O$ crystal grown in the presence of additive
Fig. 2. $CaSO_4 \cdot 2\,H_2O$ crystal grown in the absence of additive

3.2. Infrared spectral studies

A Perkin Elmer Model 283 infrared spectrophotometer was used employing KBr pellet for recording the IR spectrum (Fig. 3) of the sample grown in the presence of the additive. The strong band appearing around 1140 cm^{-1} corresponds to the asymmetric stretching vibration of SO_4^{--}. The observation of two distinct bands in the regions of stretching (3400 cm^{-1}) and bending (1650 cm^{-1}) O–H modes clearly identifies the presence of two water molecules in the crystal (BEN-DOR, MARGALITH).

3.3. Atomic absorption spectral studies

With the help of a Varian Techtron atomic absorption spectrophotometer model AA6 the percentage composition of the metals present in the crystal was determined with

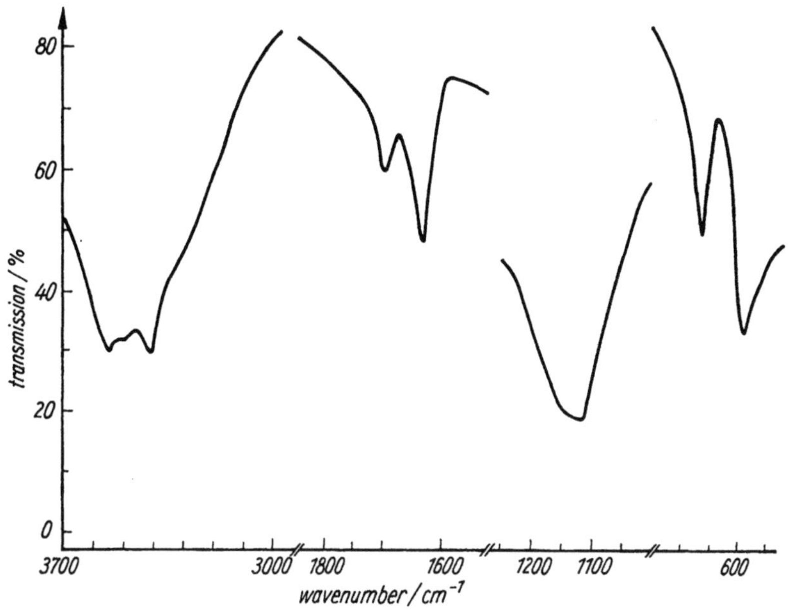

Fig. 3. Infrared Spectrum of $CaSO_4 \cdot 2\,H_2O$ grown in the presence of additive

a detection limit of 5 ppm. Accordingly, the sample was found to contain: a) calcium 21%, b) barium NIL.

3.4. Thermogravimetric studies

For thermogravimetric analysis of the sample, runs were carried out in nitrogen atmosphere on a Dupont 990 modular thermal analysis system with 951 TGA. The heating rate was maintained at 10 °C min^{-1} when the rate of flow of nitrogen was 50 cm^3 min^{-1}. From the studies on the weight loss against temperature (Fig. 4) it is clear that dehydration occurs in the range 115—165 °C. The computed value of the percentage of water in the sample is 20.97.

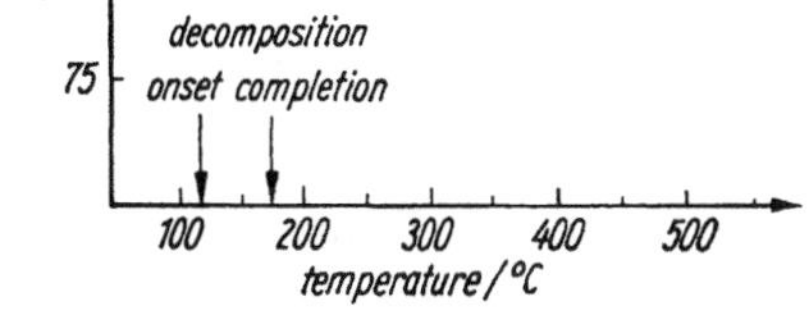

Fig. 4. The dehydration of $CaSO_4 \cdot 2\,H_2O$ grown in the presence of additive — T.G.A. data

3.5. Laser Raman studies

To record the Raman spectra of the samples SPEX Ramalog equipped with 1401 double monochromator was used. 5145 Å line of the Spectra Physics 165 Ar^+ laser was used as the excitation source. Bands due to the vibration of SO_4^{--} ion and water molecules have been assigned (Tab. 1) for both the type of crystals.

Table 1

in the presence of additive		in the absence of additive		assignment
IR cm^{-1}	Raman cm^{-1}	IR cm^{-1}	Raman cm^{-1}	
450	461	410	419	ν_2 SO_4
510	500	510	500	
600	602	600	602	
		700	676	ν_4 SO_4
1150	1140	1150	1140	ν_3 SO_4
3410	3412	3410	3412	ν_1 and ν_3 H_2O

ν_1 Symmetric stretching
ν_2 Symmetric bending
ν_3 Asymmetric stretching
ν_4 Asymmetric bending

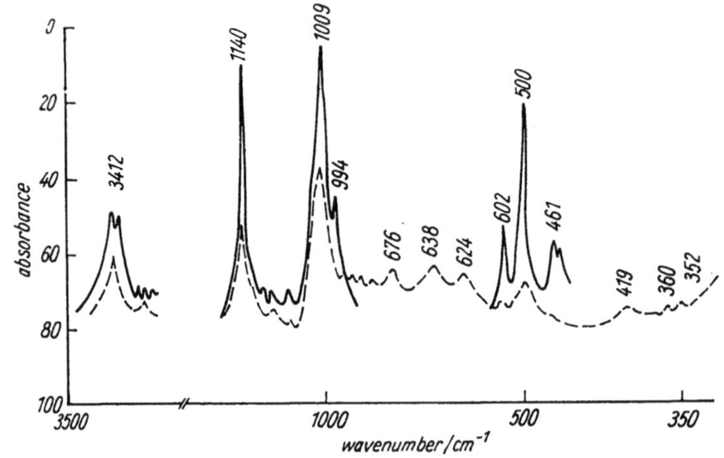

Fig. 5. (Full line) Laser Raman spectrum of the crystal grown in the presence of the additive; (dotted line) Laser Raman spectrum of the crystal grown in the absence of the additive

4. Discussion

(i) The AAS and IR studies reveal the presence of calcium and dihydrated sulphate, respectively. Barium is not incorporated in the crystal. Further from the TGA studies the crystal is confirmed to be dihydrated, and therefore it can be identified as $CaSO_4 \cdot 2\,H_2O$.

(ii) In the laser Raman spectrum of the crystal grown in the presence of the additive (Fig. 5 — full line) the bands due to SO_4^{--} appear with appreciably high intensity showing that the molecular groups are occupying their lattice sites in the crystal (Thirumavalavan et al.). The observed broadening and splitting of the bands for the sample grown in the absence of the additive (Fig. 5 — dotted line) correspond to the distorted SO_4^{--} ion. This indicates an inhomogeneous growth of the crystal.

(iii) Calcium chloride and barium chloride react separately with H_2SO_4. Due to its sparing solubility in water $BaSO_4$ precipitates readily. In this process all the sulphate radicals in the gel near the interface will be exhausted quickly. This accounts for the column of white precipitate. As the SO_4^{--} ions are absent in this region calcium and chloride ions have to diffuse slowly through this region without taking part in any reaction. Since the calcium ions experience a heavy retardation while moving through the precipitate column, the reaction with SO_4^{--} which takes place in the gel medium below the precipitate is very slow. Slow diffusion leads to perfect crystals. This explains the thin and good quality crystals obtained in the process. In the absence of the additives the crystals either displaces or traps the gel as it grows (HANKOA; NICKEL, HENISCH). The ready occurrence of reaction, in this case, gives a turbid and translucent crystal due to gel inclusion.

(iv) As the calcium ions continue to diffuse through the gel, the SO_4^{--} radicals in the vicinity of these ions will be speedily utilised to form $CaSO_4$. The loss is compensated by the diffusing sulphate ions from the column of dilute H_2SO_4. Further, any accidental leakage in the sealed region will result in an increase of density of the gel. The gel shrinks and consequently the rate of growth of the crystal is affected. The column of dilute H_2SO_4, apart from its function of compensation, plays a role in maintaining the density of the gel throughout the growth process.

Though this process has improved the harvest of crystals, the population density of the crystals is found to decline with the diffusion depth. The concentration C of the outer electrolyte diffusing into the gel decreases exponentially with the diffusion depth x. At any instant t, C is given by JOST:

$$C = \frac{S}{2} (\pi D t)^{-1/2} \exp\left\{\frac{-x^2}{4Dt}\right\}, \tag{1}$$

where D, the effective diffusion coefficient and S are constants. Any variation in t does not alter the pre-exponential term as much as it affects the exponential term. Hence $S/2\,(\pi D t)^{-1/2}$ is a constant.

Thus equation (1) implies that

$$\frac{x^2}{t} = \text{constant} \tag{2}$$

and

$$\frac{\partial C}{\partial t} \propto \frac{1}{x^2}. \tag{3}$$

One may, therefore, conclude that the concentration of the outer electrolyte is more predominant at the precipitate — gel interface than at other lower regions. C decreases rapidly with the diffusion depth and soon a stage will be reached when the concentration of the inner and outer electrolytes become equal. Nucleation occurs till this critical state appears and it ceases thereafter. This provides a satisfactory explanation for the absence of crystals at points far away from the gel-precipitate interface.

The authors wish to thank Mr. C. Muthu and Dr. M. Kanagkvel (ISRO, Trivandrum) and Dr. R. Lekshmana Sarma (Department of Chemistry, Vivekananda College, Agasteeswaram, Tamil Nadu) for their helpful suggestions and continued interest during the course of the work. One of the authots (P.K.) would like to thank Dr. M. A. Ittyachan of the Gandhiji University for introducing him in this problem and encouragements. Our thanks to Mr. V. Ramakrishnan (Department of Physics, University of Kerala, Trivandrum) for his help in connection with Raman Spectra are recorded herewith.

References

BEN-DOR, L., MARGALITH, R.: Inorganica Chimica Acta **1:1**, 51 (1967)
CHING, W., MCCARTNEY, E. R.: J. Appl. Chem. Biotechnol. **23**, 441 (1973)
HANKOA, J. I.: J. Appl. Physics **40**, 2694 (1969)
HENISCH, H. K.: Crystal Growth in Gels, Pennsylvania State University 1973
JOST, W.: Diffusion in Solids, Liquids and Gases, New York 1970
LIU, S. T., NANCOLLAS, G. H.: J. Cryst. Growth **6**, 281 (1970)
MCCARTNEY, E. R., ALEXANDER, A. E.: J. Colloid Sci. **13**, 383 (1958)
NICKEL, M. J., HENISCH, H. K.: J. Electrochem. Soc. **116**, 1258 (1969)
PATEL, A. R., BHAT, H. L.: J. Crystal Growth **12**, 288 (1972)
RAJU, K. S.: Cryst. Res. Technol. **18**, 9K, 109 (1983)
SLOVENO, M., TEZAK, G.: Colloid Polym. Sci. **254**, 900 (1976)
TADROS, MAHER, E., MAYES, IRVIN: J. Colloid and Interface Sci. **72**, 245 (1979)
THIRUMAVALAN, M., KUMAR, J., GNANAM, F. D., RAMASAMY, P.: Infra. Phy. **26**, 101 (1986)
VAN ROSMALEN, G. M., MARCHEE, W. G. J., BENNEMA, P.: J. Cryst. Growth **35**, 169 (1976)

(Received May 13, 1987)

Authors' address:

P. KUMARESON*, Prof. S. DEVANARAYANAN
Department of Physics, University of Kerala
Kariavattom
Trivandrum — 695581
India

* *Present address*:
Department of Physics, Vivekananda College
Agasteeswaram — 629701
Tamil Nadu
India

Cryst. Res. Technol.	22	1987	12	1459–1470

D. I. ISMAILOV, YU. G. ASADOV, G. B. GASYMOV

Institute of Physics of Academy of Science of Azerb. SSR, Baku, USSR

Structural Phase Transitions in $Cu_{2-x}Fe_yS$ Crystals

Crystals with $Cu_{2-x}Fe_yS$ compositions ($x = 0.30, 0.25, 0.20, 0.15$; $y = 0.05$) have been synthetized and obtained by the method of slow cooling. X-ray diffractometric studies have shown that at room temperature the above crystals consist of an orthorhombic phase (anilite), a monoclinic phase (djurlite), and their metastable high-temperature face-centered cubic phases $(FCC)_1$ and $(FCC)_2$, respectively. A crystallographic relation has been found between the lattices of the orthorhombic, monoclinic and $(FCC)_1$ phase. It is shown that all these phases transform in the final analysis to a single high-temperature fcc phase. All the transitions are reversible, i.e. in the course of inverse cooling to room temperature, the diffraction patterns are restored completely in reverse order.

According to GEZALOV et al., a clearly defined polymorphism and a statistical distribution of copper atoms among individual positions of a high-temperature face-centered cubic (fcc) lattice, the total multiplicity of which greatly exceeds the number of atoms in a cell, are typical of the $Cu_{2-x}S$ ($0.25 \geqq x \geqq 0$) system.

At room temperature the crystals of all the non-stoichiometric $Cu_{2-x}S$ compounds are two-phase, i.e. they consist of anilite ($Cu_{1.75}S$) and djurlite ($Cu_{1.96}S$). When djurlite is synthetized it is accompanied by a chalcocite mixture (Cu_2S). A thorough analysis of diffraction patterns, performed by us, has shown that at room temperature a metastable high-temperature fcc modification exists in $Cu_{2-x}S$ ($x = 0.15, 0.20, 0.25$) crystals along with an orthorhombic (anilite) and a monoclinic (djurlite) modification. The structural phase transitions taking place in the crystals of the $Cu_{2-x}S$ ($0.25 > x < 0$) system are summarized in Diagram 1. Anilite crystallizes in the space group Pnma with the unit cell parameters $a = 7.89$, $b = 7.84$, $c = 11.01$ Å, and $Z = 4$ (KOTO, MORIMOTO), whereas djurlite crystallizes in $P\,2_1/n$ with $a = 26.897$, $b = 15.745$, $c = 13.565$ Å, $\beta = 90.13°$, and $Z = 8$ $Cu_{31}S_{16}$ (EVANS).

As a rule, the orthorhombic-to-monoclinic modification ratio in the above compositions is defined by the value of x. At $x = 0.25$ the monoclinic modification is at its minimum in the specimen and grows with decreasing deficit of copper atoms, showing a prevalence in $Cu_{1.90}S$ crystals.

The present report generalizes the results of the effect of partial replacement of copper atoms by iron atoms on the regularities of polymorphic transformations in $Cu_{2-x}S$ crystals. For this purpose, $Cu_{2-x}Fe_yS$ ($x = 0.30, 0.25, 0.20, 0.15$; $y = 0.05$) crystals were synthetized and obtained by the method of slow cooling.

An X-ray diffractometric temperature-investigation was conducted in a "DRON-1.5"-type diffractometer with a temperature attachment operating in the range from 293 to 2273 K.

Experimental part

1. $Cu_{1.70}Fe_{0.05}S$

Eighteen distinct diffraction reflections have been obtained from one crystal at room temperature in the ($10° \leqq 2\theta \leqq 100°$) angle range. The reflections can be indexed on

Diagram 1

Diagram of structural transformations in $Cu_{2-x}S$ compositions

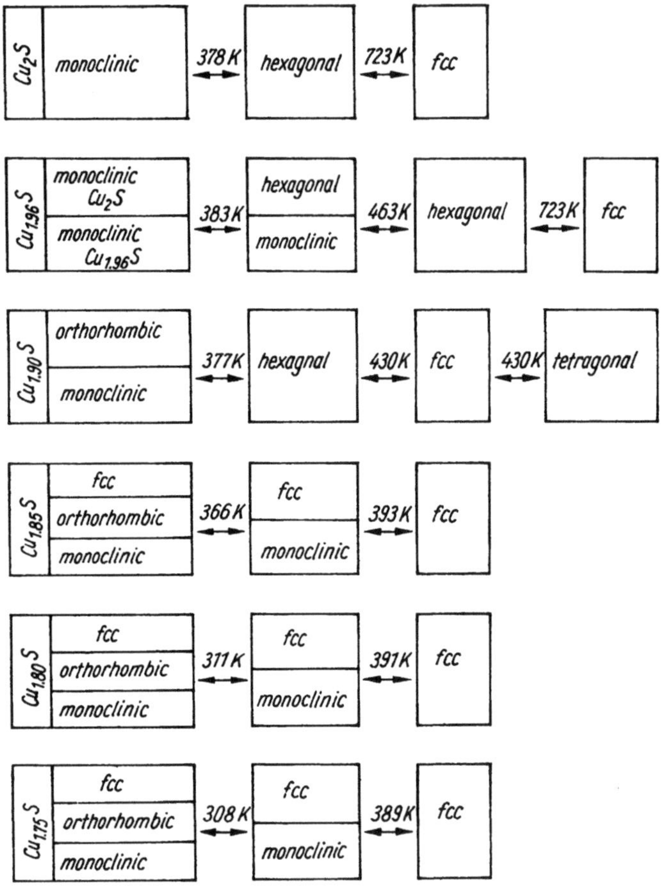

the basis of the orthorhombic (anilite), monoclinic (djurlite) modification as well as their metastable high-temperature $(FCC)_1$ and $(FCC)_2$ modifications differing from one another by the unit cell parameters and metal-atom concentrations. Table 1 contains the results of calculations based on a powder pattern and a diffraction pattern taken with the use of a CoK and CuK radiation, respectively. This table shows that, the diffraction reflections from the (111), (222) and (333)-type planes of an $(FCC)_1$-modification lattice are superimposed by the diffraction reflections from the (202) (404), and (606)-type planes of an orthorhombic-modification lattice and the reflections from the (333), (666), and (999)-type planes of a monoclinic-modification lattice. Upon recording the diffraction reflections at room temperature a heater was turned-on without disturbing the original crystal orientation, and only seven diffraction reflections, four of them belonging to $(FCC)_1$, one to $(FCC)_2$ and the other two to the monoclinic modification, were recorded in the original range of angles at 343 K (Tab. 2).

If the heating of the specimen is continued, the diffraction reflections from the lattices of the monoclinic and fcc modifications disappear at 393 K, and only three reflections from planes (111), (222), and (333), belonging to the high-temperature

Table 1
Interplanar spacings of $Cu_{1.70}Fe_{0.05}S$

Nos	d_p/Å	I/I_0	d_d/Å	orthorhombic		monoclinic		$(FCC)_1$		$(FCC)_2$	
				d_t/Å	*hkl*	d_t/Å	*hkl*	d_t/Å	*hkl*	d_t/Å	*hkl*
1	–	–	3.964	–	–	3.9646	232	–	–	–	–
2	–	–	3.201	3.2066	202	3.1982	333	3.2038	111	–	–
3	3.170	6	–	–	–	3.1695	304	–	–	–	–
4	3.044	2	–	–	–	3.0475	151	–	–	–	–
5	2.993	2	–	–	–	2.9903	251	–	–	–	–
6	2.749	7	–	–	–	2.7207	352	–	–	–	–
7	–	–	2.678	2.6870	203	–	–	–	–	–	–
8	2.519	1	2.524	–	–	2.5228	244	–	–	–	–
9	2.291	2	2.285	–	–	2.2738	254	–	–	–	–
10	2.167	8	–	2.1661	124	2.1666	345	–	–	–	–
11	–	–	2.105	–	–	2.1083	272	–	–	–	–
12	–	–	2.002	–	–	2.0027	355	–	–	1.9971	220
13	1.958	10	–	1.9600	040	1.9605	046	–	–	–	–
14	1.840	2	1.829	1.8280	331	–	–	–	–	–	–
15	–	–	1.780	–	–	1.7809	337	–	–	–	–
16	1.748	1	–	1.7426	116	1.7424	090	–	–	–	–
17	1.670	9	1.678	1.6781	422	1.6806	292	1.6731	311	–	–
18	–	–	1.632	1.6273	216	1.6344	418	–	–	1.6304	222
19	1.605	1	1.595	1.6033	404	1.5991	666	1.6019	222	–	–
					044						
20	1.451	2	1.458	1.4571	250	1.4568	458	–	–	–	–
21	1.393	3	–	1.3903	440	1.3927	468	–	–	–	–
22	1.312	2	–	1.3150	600	1.3121	0.12.0	–	–	–	–
23	1.257	1	1.265	1.2644	353	1.2651	4.12.2	–	–	–	–
24	1.134	7	1.145	1.1446	362	1.1443	4.7.10	–	–	–	–
25	–	–	1.118	1.1188	633	1.1188	4.12.6	–	–	–	–
26	–	–	1.086	1.0857	356	1.0865	5.14.2	–	–	1.0869	333
27	1.069	6	1.070	1.0689	606	1.0661	999	1.0679	333	–	–
					257		5.14.3				
28	0.985	5	–	0.9863	800	0.9841	0.16.0	–	–	–	–
29	0.942	5	–	0.9462	654	0.9451	0.16.4	–	–	–	–

Table 2
Interplanar spacings of $Cu_{1.70}Fe_{0.05}S$

Nos	d_d/Å	monoclinic		$(FCC)_1$		$(FCC)_2$		T/K
		d_t/Å	*hkl*	d_t/Å	*hkl*	d_t/Å	*hkl*	
1	3.224	–	–	3.2252	111	–	–	
2	2.108	2.1083	272	–	–	–	–	
3	1.687	1.6806	292	1.6843	311	–	–	
4	1.610	–	–	1.6125	222	–	–	373
5	1.126	1.1257	3.12.6	–	–	–	–	
6	1.088	1.0865	5.14.2	–	–	1.0871	333	
7	1.075	–	–	1.0750	333	–	–	
1	3.229			3.2309	111			393
2	1.615			1.6154	222			
3	1.077			1.0769	333			

d_p interplanar spacing calculated from powder pattern.
d_d interplanar spacing calculated from diffraction pattern.
d_t interplanar spacing calculated theoretically.

Table 3
Interplanar spacings of $Cu_{1.75}F3e_{0.05}S$

Nos	d_p/Å	I/I_0	d_d/Å	orthorhombic		monoclinic		$(FCC)_1$		$(FCC)_2$	
				d_t/Å	*hkl*	d_t/Å	*hkl*	d_t/Å	*hkl*	d_t/Å	*hkl*
1	3.945	4	3.945	3.9450	200	3.9363	040	—	—	—	—
2	3.876	4	3.878	—	—	3.8788	123	—	—	—	—
3	3.258	1	—	—	—	—	—	—	—	3.2608	111
4	3.196	3	3.197	3.2066	202 022	3.1982	333	3.2078	111	—	—
5	3.149	7	—	—	—	3.1490	050	—	—	—	—
6	3.023	5	—	—	—	3.0252	404	—	—	—	—
7	2.713	9	2.712	—	—	2.7130	005 334	—	—	—	—
8	2.674	3	2.675	—	—	2.6736	015	—	—	—	—
9	2.513	1	—	—	—	2.5140	405	—	—	—	—
10	2.276	7	—	2.2713	312	2.2712	435	—	—	—	—
11	2.145	6	—	2.1423	321	2.1444	406 226	—	—	—	—
12	1.964	10	—	1.9725	400	1.9681	080	—	—	—	—
13	1.872	1	—	1.8734	233	1.8716	282	—	—	—	—
14	1.827	2	1.828	1.8280	331	1.8239	424	—	—	—	—
15	1.778	1	1.779	—	—	1.7791	565	—	—	—	—
16	1.760	1	1.759	1.7568	332	1.7540	437	—	—	—	—
17	1.710	2	—	1.7109	324	1.7104	583	—	—	—	—
18	1.672	7	1.672	1.6723	242	1.6723	218	1.6750	311	—	—
19	1.603	1	1.604	1.6033	404	1.5991	666 393	1.6039	222	—	—
20	1.554	1	—	1.5558	341	1.5548	094 148	—	—	—	—
21	1.393	3	—	1.3903	440	1.3927	468	—	—	—	—
22	1.315	3	—	1.3150	600	1.3121	0.12.0				
23	1.133	4	—	1.1357	624	1.1348	0.12.6	1.1341	422	—	—
24	1.086	1	1.085	1.0840	461 625	1.0833	3.14.3	—	—	1.0869	333
25	1.069	4	1.070	1.0689	606 257	1.0661	999 5.14.3	1.0693	333	—	—
26	0.982	1	—	0.9863	800 801	0.9841	0.16.0	0.9821	440	—	—
27	0.955	1	—	0.9525	803	0.9516	0.16.3	—	—	—	—
28	0.941	1	—	0.9423	822	0.9451	0.16.4	—	—	—	—

Table 4
Interplanar spacings of $Cu_{1.75}Fe_{0.05}S$

Nos	d_d/Å	fcc		lattice spacing/Å	T/K
		d_t/Å	*hkl*		
1	3.224	3.2130	111		
2	1.612	1.6065	222	$a = 5.586$	353
3	1.075	1.0750	333		

$(FCC)_1$ modification, are recorded in the original range of angles (Tab. 2). As seen from Tables 1 and 2, the unit cell parameter a of the $(FCC)_1$ modification varies from 5.549 to 5.596 Å in the temperature range from 293 to 393 K, due to a gradual dissolution of $(FCC)_2$ and transformation of the monoclinic modification.

The appearance of an $(FCC)_2$ modification and its dissolution with increasing temperature in the case of an $(FCC)_1$ modification is dealt with below.

Table 5
Interplanar spacings of $Cu_{1.80}Fe_{0.05}S$

Nos	d_p/Å	I/I_0	d_d/Å	orthorhombic		monoclinic		$(FCC)_1$		$(FCC)_2$	
				d_t/Å	*hkl*	d_t/Å	*hkl*	d_t/Å	*hkl*	d_t/Å	*hkl*
1	–	–	3.970	–	–	3.9646	232	–	–	–	–
2	–	–	3.262	–	–	–	–	–	–	3.2619	111
3	–	–	3.205	3.2066	202	3.1982	333	3.2112	111	–	–
4	–	–	3.161	–	–	3.1695	304	–	–	–	–
5	3.149	4	–	–	–	3.1490	050	–	–	–	–
6	2.967	4	–	2.9680	212	2.9689	043	–	–	–	–
7	2.689	9	2.681	2.6791	203	–	–	–	–	–	–
8	2.494	2	–	2.4934	310	–	–	–	–	–	–
9	–	–	2.289	–	–	2.2738	254	–	–	–	–
10	2.250	8	–	2.2526	024	2.2525	106	–	–	–	–
11	2.130	6	–	2.1372	231	2.1349	072	–	–	–	–
12	–	–	2.108	–	–	2.1083	272	–	–	–	–
13	–	–	2.003	–	–	2.0027	355	–	–	–	–
14	1.968	10	–	1.9725	400	1.9681	080	–	–	1.9977	220
15	1.942	3	–	1.9416	401	1.9426	181	–	–	–	–
16	1.874	1	–	1.8768	323	1.8745	074	–	–	–	–
17	1.762	1	1.784	1.7867	016	1.7809	337	–	–	–	–
18	1.710	4	–	1.7109	324	1.7104	583	–	–	–	–
19	1.662	5	1.677	1.6781	422	1.6806	292	1.6770	311	–	–
20	–	–	1.631	1.6276	216	1.6344	418	–	–	1.6309	222
21	–	–	1.608	1.6033	404	1.5991	666	1.6056	222	–	–
22	1.594	1	1.581	1.5835	243	1.5818	285	–	–	–	–
23	1.399	5	–	1.3903	440	1.3996	249	–	–	–	–
24	1.313	3	–	1.3150	600	1.3121	0.12.0	–	–	–	–
25	–	–	1.280	1.2804	161	1.2803	6.2.10	–	–	–	–
26	1.253	1	–	1.2552	162	1.2521	079	–	–	–	–
27	1.188	1	–	1.1865	604	1.1876	2.13.2	–	–	–	–
28	–	–	1.147	1.1446	362	1.1470	2.11.7	–	–	–	–
29	1.127	2	–	1.1271	700	1.1246	0.14.0	–	–	–	–
30	–	–	1.088	1.0879	536	1.0880	7.12.6	–	–	1.0872	333
31	–	–	1.071	1.0689	606	1.0661	999	1.0704	333	–	–
					642		1.9.10				
32	–	–	1.055	1.0548	166	1.0551	11.13.3	–	–	–	–

2. $Cu_{1.75}Fe_{0.50}S$

The indexing of the powder and diffraction patterns recorded with a specimen of crystalline $Cu_{1.75}Fe_{0.05}S$ and presented in Table 3 has shown that at room temperature these crystals also consist of an orthorhombic and a monoclinic modification and their metastable $(FCC)_1$ and $(FCC)_2$ modifications.

Only three of the twelve diffraction reflections recorded at room temperature remain at 353 K, namely those reflected from the (111), (222), and (333)-type planes and belonging to the $(FCC)_1$ modification (see Tab. 4). In this case the $(FCC)_2$ modification also dissolves gradually as the temperature is increased, and as a result a single fcc modification remains with a unit cell parameter a at $\Delta T = T_{\text{transf.}} - T = 54$ K ranging from 5.556 to 5.584 Å.

3. $Cu_{1.80}Fe_{0.05}S$

As seen from Table 5, at room temperature $Cu_{1.80}Fe_{0.05}S$ crystals also consist of an orthorhombic and a monoclinic modification and their metastable $(FCC)_1$ and $(FCC)_2$ modifications, respectively, with the unit cell parameters $a_1 = 5.562$ Å and $a_2 = 5.649$ Å. At a temperature of 342 K the orthorhombic miodfication transforms to an $(FCC)_1$

Table 6
Interplanar spacings of $Cu_{1.80}Fe_{0.05}S$

Nos	d_d/Å	monoclinic		$(FCC)_1$		$(FCC)_2$		lattice spacing Å	T/K
		d_t/Å	hkl	d_t/Å	hkl	d_t/Å	hkl		
1	3.263	–	–	–	–	3.2637	111		
2	3.225	3.1982	333	3.2251	111	–	–		
3	2.774	2.7703	234	–	–	–	–		
4	2.103	2.1083	272	–	–	–	–		
5	1.632	1.6344	418	–	–	1.6316	222	a_1 = 5.586	
6	1.612	1.5991	666	1.6125	222	–	–		373
7	1.581	1.5818	285	–	–	–	–	a_2 = 5.652	
			276						
8	1.088	1.0880	7.11.7	–	–	1.0877	333		
9	1.075	1.0711	999	1.0750	333	–	–		
10	1.055	1.0551	11.13.3	–	–	–	–		
1	3.234			3.2343	111	–	–		
2	1.617			1.6172	222	–	–	a_1 = 5.602	
3	1.088			–	–	1.0881	333		423
4	1.078			1.0781	333	–	–	a_2 = 5.654	
1	3.255			3.2517	111				
2	1.625			1.6258	222			a = 5.632	473
3	1.083			1.0839	333				

modification, and as this takes place, only 10 of 18 diffraction reflections are recorded in the original range of angles, as shown in Table 6. Four of the ten reflections belong to the monoclinic modification, three of them represent reflections from planes (111), (222), and (333) of an $(FCC)_1$ lattice and the other three are reflections from planes (111), (222), and (333) of an $(FCC)_2$-modification lattice.

At 386 K the monoclinic modification turns to a $(FCC)_2$ modification. In this case only four of the ten diffraction reflections presented in Table 6 are recorded, three of them being reflections from planes (111), (222), and (333) of an $(FCC)_1$ lattice and one of them representing a reflection from plane (333) of an $(FCC)_2$-modification lattice.

If one continues to heat the specimen above 438 K, the reflection from plane (333), belonging to the $(FCC)_2$ modification, disappears whereas the reflections from planes (111), (222), and (333) of an $(FCC)_1$-modification lattice are displaced towards smaller angles (see Tab. 6). A single fcc modification produced by mixing two modifications, namely $(FCC)_1$ and $(FCC)_2$, has a lattice spacing $a = 5.632$ Å (see Fig. 3).

4. $Cu_{1.85}Fe_{0.05}S$

The powder-pattern and diffraction-pattern data presented in Table 7 show that at room temperature $Cu_{1.85}Fe_{0.05}S$ crystals consist of an orthorhombic, a monoclinic, and their metastable high-temperature $(FCC)_1$ and $(FCC)_2$ modifications.

At a temperature of $T_{transf.} > T = 385$ K the orthorhombic modification transforms to an $(FCC)_1$ modification, i.e. only six of the fifteen diffraction reflections recorded at room temperature remain (see Tab. 8). Four of these reflections belong to a monoclinic modification, whereas two of them are reflections from planes (111) and (222) of an $(FCC)_1$-modification lattice.

If heating is continued, all of the four reflections from the monoclinic modification disappear at 573 K, and four reflections from the $(FCC)_1$ modification are recorded in the original range of angles (Tab. 8). The monoclinic-to-fcc modification transformation temperature is equal to 542 ± 2 K.

Table 7
Interplanar spacings of $Cu_{1.85}Fe_{0.05}S$

Nos	d_p/Å	I/I_0	d_d/Å	orthorhombic		monoclinic		$(FCC)_1$		$(FCC)_2$	
				d_t/Å	*hkl*	d_t/Å	*hkl*	d_t/Å	*hkl*	d_t/Å	*hkl*
1	–	–	3.366	3.3563	211	3.3637	104	–	–	–	–
2	–	–	3.256	–	–	–	–	–	–	3.2608	111
3	3.214	3	–	3.2066	202	3.1982	333	3.2147	111	–	–
4	–	–	3.146	–	–	3.1490	050	–	–	–	–
5	2.966	3	–	–	–	2.9689	042	–	–		
6	2.772	2	2.763	2.7525	004	–	–	–	–	–	–
7	2.713	8	–	–	–	2.7130	334 005	–	–	–	–
8	2.524	3	2.533	2.5369	123	2.5371	253 261				
9	2.494	1	–	2.4934	310	–	–	–	–	–	–
10	2.259	8	–	2.2526	024	2.2525	106	–	–	–	–
11	2.223	4	–	–	–	2.2224	1.12.4	–	–	–	–
12	–	–	2.099	2.0625	313	2.0764	036 563	–	–	–	–
13	1.973	10	–	1.9600	040	1.9605	046	–	–	–	–
14	1.861	9	1.879	1.8769	323	1.8797	317	–	–	–	–
15	1.750	6	–	1.7426	116	1.7494	090	–	–	–	–
16	1.681	5	1.686	1.6839	035	1.6859	018 118	1.6788	311	–	–
17	1.609	7	1.581	1.6033	404 044	1.5991	666 285	1.6073	222	–	–
18	1.393	4	–	1.3903	440	1.3927	468	–	–	–	–
19	1.318	6	1.327	1.3284	137	1.3269	479	–	–	–	–
20	1.261	6	1.264	1.2644	353	1.2645	1.11.5	–	–	–	–
21	1.230	1	–	1.2310	063	–	–	–	–	–	–
22	1.170	4	1.170	1.1702	360	1.1709	3.12.5	–	–	–	–
23	1.126	4	1.126	1.1263	048	1.1257	3.12.6	–	–	–	–
24	1.086	1	1.087	1.0870	172	1.0873	6.14.1	–	–	1.0869	333
25	–	–	1.055	1.0548	166	1.0551	11.13.3	–	–	–	–

Table 8
Interplanar spacings of $Cu_{1.85}Fe_{0.05}S$

Nos	d_d/Å	monoclinic		fcc		lattice spacing Å	*T*/K
		d_t/Å	*hkl*	d_t/Å	*hkl*		
1	3.379	3.3637	104	–	–		
2	3.236	–	–	3.2368	111		
3	1.695	1.6859	018	–	–		
4	1.619	–	–	1.6184	222	$a = 5.607$	473
5	1.133	1.1257	3.12.6	–	–		
6	1.084	1.0833	3.14.3	–	–		
1	3.271			3.2725	111		573
2	2.834			2.8340	200	$a = 5.668$	
3	1.634			1.6262	222		
4	1.091			1.0908	333		

As seen from Table 8, the diffraction reflections belonging to the $(FCC)_2$ modification disappear as with the above compositions, i.e. as the temperature is increased, this modification dissolves, and a single modification, namely $(FCC)_1$ remains. As with the other compositions discussed above, a growth of the unit cell parameter of this modification from $a = 5.568$ Å at $T = 293$ K to $a = 5.665$ Å at $T = 573$ K is observed.

On cooling to room temperature, the specimen regains its initial state, i.e. as the crystals are cooled the diffraction patterns are restored completely in reverse order.

Discussion of results

The process involving the synthesis and growth of crystals with non-stoichiometric $Cu_{2-x}Fe_yS$ compositions ($x = 0.30$, 0.25, 0.20, 0.15; $y = 0.05$) takes place at temperatures higher than those of structural transformations. That is why in the course of synthetizing and growing crystals the first crystals to be formed and to grow are those of the high-temperature fcc modification, and the first cycle of transformations takes place when the specimen is cooled from crystallization to room temperature.

At room temperature $Cu_{2-x}Fe_yS$ crystals differ from $Cu_{2-x}S$ ($x = 0.25$, 0.20, 0.15, 0.10) crystals in that in addition to an orthorhombic (anilite), a monoclinic (djurlite) and a metastable fcc modification they also contain a second fcc modification differing in metal-atom concentration and hence the unit cell parameter. A comparison of the unit cell parameters of the $(FCC)_1$ and $(FCC)_2$ modifications of $Cu_{2-x}Fe_yS$ compositions with the unit cell parameters of the fcc modifications of anilite and djurlite has shown that the unit cell parameter of the $(FCC)_1$ modification coincides with the unit cell parameter of the fcc modification of anilite ($Cu_{1.75}S$), whereas the $(FCC)_2$ parameter coincides with the unit cell parameter of the fcc modification of djurlite ($Cu_{1.96}S$).

Diagram 2

Diagram of structural transformations in $Cu_{2-x}Fe_{0.05}S$ compositions

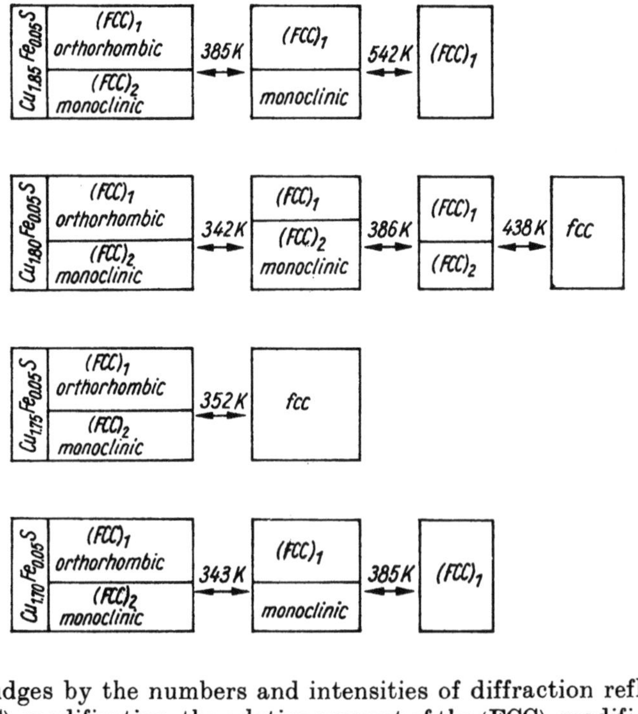

If one judges by the numbers and intensities of diffraction reflections belonging to the $(FCC)_2$ modification, the relative amount of the $(FCC)_2$ modification is minimum and unstable in the above compositions with the exception of $Cu_{1.80}Fe_{0.05}S$.

As seen from the diagram of structural transformations in $Cu_{2-x}Fe_yS$ crystals, the $(FCC)_2$ modification merges gradually with the $(FCC)_1$ modification as the temperature is increased. However, with a $Cu_{1.80}Fe_{0.05}S$ crystal, subsequent to orthorhombic → $(FCC)_1$ and monoclinic → $(FCC)_2$ modification transformations the specimen consists of an $(FCC)_1$ and an $(FCC)_2$ modification. Under annealing at 438 K, these modifications mix together to form a single fcc modification with the unit cell parameter $a = 5.632$ Å, where $a_1 < a < a_2$, as might be expected.

In the fcc modifications of the above compositions the lattice skeleton belongs to sulphur atoms, and the very mobile copper atoms Cu^{1+} and Cu^{2+} are distributed statistically in the lattice and occupy the tetrahedral and octahedral voids formed between two neighbouring sulphur-atom layers. In as much as the ionic radii of bivalent copper and iron ions are equal to 0.80 Å, the bivalent iron ions replace isomorphously the bivalent copper ions in the sulphur-atom octahedron centres. In this case the electrostatic balance of the crystal lattice is not disturbed either.

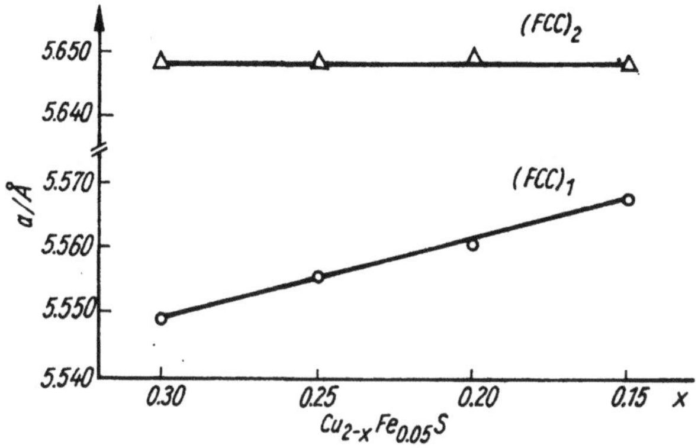

Fig. 1. Dependences of $(FCC)_1$ and $(FCC)_2$ lattice modification parameters on the $Cu_{2-x}F_{0.05}S$ composition

Upon cooling, the single high-temperature fcc modifications of $Cu_{2-x}Fe_yS$ crystals lose a part of metal atoms from the alternating sulphur-atom layers, and thus the layers enriched with metal atoms form $(FCC)_2$ modifications, whereas the layers poor in metal atoms form $(FCC)_1$ modifications. At room temperature the unit cell parameter for the $(FCC)_2$ modification in crystal with $Cu_{2-x}Fe_yS$ compositions is the same and amounts to 5.648 Å. This suggests that irrespective of the crystal composition the $(FCC)_2$ modifications — metastable at room temperature — retain only a certain number of metal atoms. The parameter of an $(FCC)_1$-modification lattice, on the contrary, grows linearly depending on the composition, as seen from Figure 1.

Figure 2 depicts the temperatures of transformation of an orthorhombic and a monoclinic modification to an fcc modification, depending on the composition of the $Cu_{2-x}Fe_yS$ crystal. As may be seen, in all the crystals, except $Cu_{1.75}Fe_{0.05}S$ where the orthorhombic and monoclinic modifications transform simultaneously to a single fcc modification, the temperatures of transformation of an orthorhombic to an $(FCC)_1$ modification are lower than the temperatures of transformation of a monoclinic to an $(FCC)_2$ modification. It is also seen that the orthorhombic fcc modification and monoclinic fcc modification transformation curves are symbathic and the transformation temperature increases with decreasing x.

Figure 3 gives the temperature dependence of the unit cell parameter of the $(FCC)_1$ modification of $Cu_{2-x}Fe_yS$ compositions, where an increase of the unit cell parameter

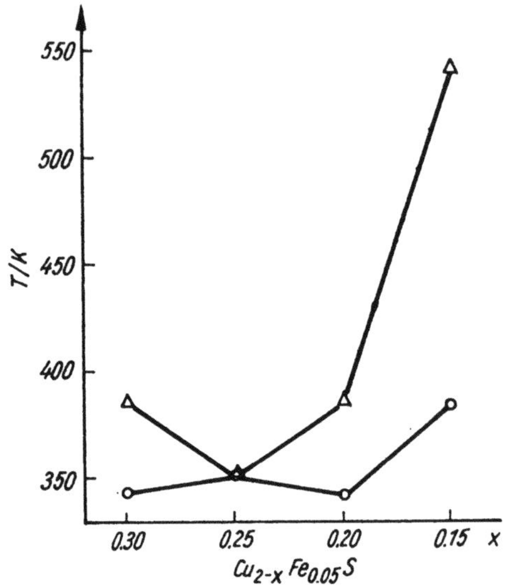

Fig. 2. Dependences of temperatures of transformations of an orthorhombic and a monoclinic modification to a fcc modification on the $Cu_{2-x}Fe_{0.05}S$ composition. △ monoclinic → fcc, ○ orthorhombic → fcc

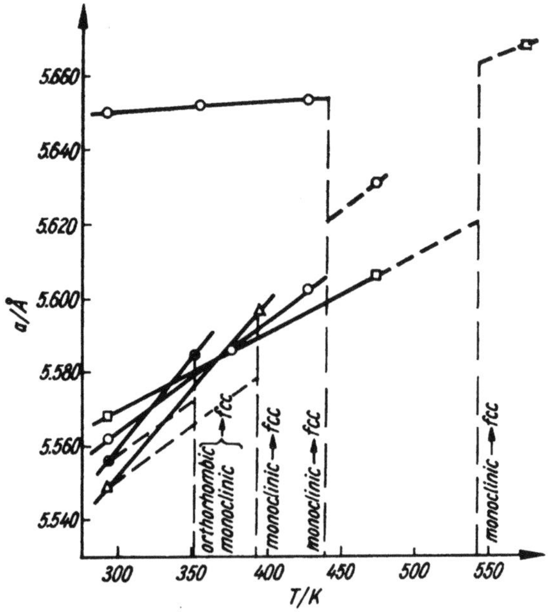

Fig. 3. Temperature dependence of unit cell parameters of $(FCC)_1$ modifications of $Cu_{2-x}Fe_{0.05}S$ compositions. △ $Cu_{1.70}Fe_{0.05}S$, ● $Cu_{1.75}Fe_{0.05}S$, ○ $Cu_{1.80}Fe_{0.05}S$, □ $Cu_{1.85}Fe_{0.05}S$

of the $(FCC)_1$ modification is due to two simultaneous processes, namely a) gradual dissolution of the djurlite fcc modification existing at room temperature, and b) a temperature process. In this case it is practically impossible to separate one of the processes from the other. But if one takes into account the fact that in the $Cu_{1.80}Fe_{0.05}S$ composition the $(FCC)_2$ modification preserves its individual nature up to a temperature higher than the transformation temperature of the monoclinic modification, the growth of the unit cell parameter of the $(FCC)_1$ modification of this composition may be attributed to the temperature increase. Then, on the basis of this fact, one can draw lines from the initial parameter values of the $(FCC)_1$ modification for the other compositions separately (these lines being parallel to the parameter growth line of the $(FCC)_1$ modification of $Cu_{1.80}Fe_{0.05}S$) to the transformation temperature of the monoclinic modification, and it will be possible to attribute them to the temperature part of the growth of parameters. In Figure 3 these lines are shown as dashed lines. Upon doing so, everything falls into place, i.e. as the monoclinic modification transforms to the fcc modification the unit cell parameter of the latter varies in a jump in all the compositions.

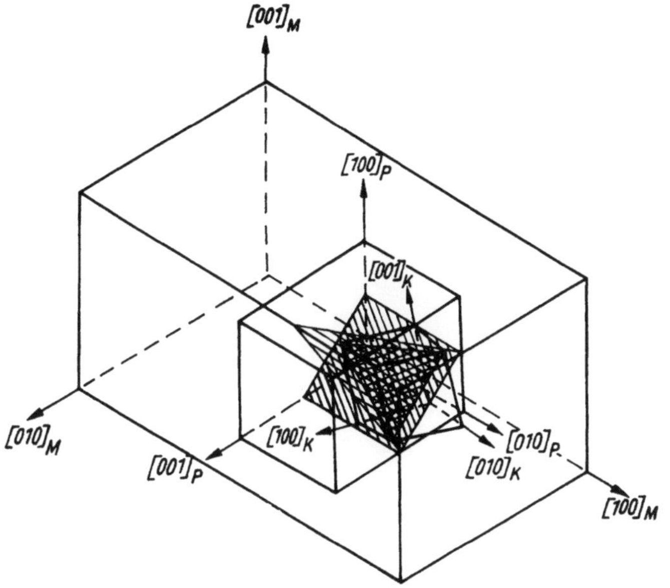

Fig. 4. Crystallographic relation between the lattices of $(FCC)_1$, orthorhombic and monoclinic modifications

In $Cu_{2-x}Fe_yS$ ($x = 0.15, 0.20, 0.25, 0.30$; $y = 0.05$) crystals, as in $Cu_{2-x}S$ ($x = 0.10, 0.15, 0.20, 0.25$) crystals, the diffraction reflections from the (111), (222), and (333)-type planes of an $(FCC)_1$-modification lattice are superimposed by reflections from the (202), (404), and (606)-type planes of an orthorhombic lattice and reflections from the (333), (666), and (999)-type planes of a monoclinic-modification lattice. Besides, the (h00)-type planes of an orthorhombic-modification lattice are parallel to the (0k0)-type planes of a monoclinic-modification lattice.

Figure 4 shows the interrelation between an $(FCC)_1$, an orthorhombic, and a monoclinic lattice, where plane (111) of a cubic lattice is superimposed by plane (202) of an orthorhombic and plane (333) of a monoclinic lattice. In the course of reversible transformations of orthorhombic and monoclinic modifications in these crystals the cubic lattice plays the part of a matrix crystal, as with an epitaxial growth, owing to their metastable high-temperature fcc modifications existing at room temperature.

A comparison between Diagrams 1 and 2 shows illustratively that a partial replacement of copper atoms by iron atoms in $Cu_{2-x}S$ compositions ($x = 0.25$, 0.20, 0.15, 0.10) leads to variations in the temperatures of structural transformations and to the appearance of separate fcc modifications of anilite and djurlite at room temperature.

References

EVANS, Jr. H. T.: Z. Krist. **150**, 299 (1979)

GEZALOV, M. A., G. B. GASYMOV, Yu. G. ASADOV, G. G. GUSEINOV, N. V. BELOV: Doklady AN SSSR **239**, 846 (1978)

KOTO, K., MORIMOTO, N.: Acta Cryst. **B26**, 915 (1970)

(Received 1) November 26, 1985;
2) May 14, 1987)

Authors' address:

D. I. ISMAILOV, YU. G. ASADOV, G. B. GASYMOV
Institute of Physics, Azerb. SSR Academy of Sciences
pr. Narimanova, 33
Baku — 370143, USSR

Cryst. Res. Technol.	22	1987	12	1471–1475

A. J. HORODECKI

Institute of Experimental Physics
University of Gdańsk, Poland

Semistatistically Defected Wurtzite Crystal Containing Pairs of Metal-Non-metal Vacancy

Basing upon the wurtzite structure the unit cell of a semistatistically defected crystal is defined as containing four unit cells of an ideal crystal and one pair of metal and non-metal vacancy. It is shown that all of possible coordinates of these vacancies lead only to three different squares of structure amplitude. The results are used for discussion of peculiarities of X-ray pattern of CdS thin films.

Auf der Basis der Wurtzitstruktur wird die Elementarzelle eines semistatistisch gestörten Kristalls als aus vier Idealkristall-Elementarzellen und einem Cadmium–Schwefel-Leerstellenpaar bestehend definiert. Es wird gezeigt, daß alle möglichen Koordinaten dieser Leerstellen nur zu drei verschiedenen Quadraten der Strukturamplitude führen. Die Ergebnisse werden für die Diskussion der Besonderheiten des Röntgendiagramms von CdS-Dünnschichten genutzt.

1. Introduction

In the previous work the peculiarities of growth of pyroelectric compound film were discussed (HORODECKI). Besides the inclination to the textured growth the origin of pairs of metal-non-metal vacancy (VP-hereafter) was enlightened. These suppositions were preliminary confirmed for the case of the vapour deposited CdS thin films (HORODECKI et al.). In spite of the well developed statistical method of investigation of point defects, little is known about the independence of physical properties on the spatial distribution of the latter. Indeed, the experimental methods cannot indirectly distinguish the coordinates of vacancies. For example, the analysis of X-ray integral intensities can give the valuable information about the composition of substitutional solids (SCHMIDT, PAHLE). Nevertheless, being extended on the case of "substitutional" vacancies it gives nothing but their average number.

Here, we investigate the periodical arrangements of different VP in wurtzite structure. By calculation of the corresponding squares of structure amplitude the concept of semistatistically defected crystal (SW-hereafter) is confirmed in terms of spatial distribution of VP. The results obtained are discussed together with experimental intensities of X-ray reflections of CdS thin films.

2. Spatial distribution of VP

Let us assume the ideal wurtzite crystal (IW-hereafter) divided into cells with edges of $2a_0$, $2a_0$ and $1c_0$ respectively, where a_0 and c_0 are the edges of common unit cell of IW. To obtain the base for our considerations we postulate this cell containing exactly one VP. By virtue of this postulate the question of spatial distribution of VP is reduced to the analysis of the intracellular location both metal as well non-metal vacancy. In fact, we define in the above manner SW but in order to calculate its structure amplitude, F_{hkl} we first analyse the above cell without of any VP. It contains, of course, four unit cells of IW i.e. eight atoms of metal and nonmetal (M-, and X-hereafter), respectively. The collection of their coordinates can be written as follows

$$\begin{aligned}\Gamma = M\{&[(0,0,0),(12,0,0),(0,12,0),(12,12,0)];(4,8,12)[(0,0,0),\\&(12,0,0),(0,12,0),(12,12,0)]\} + X\{(0,0,9)[(0,0,0),(12,0,0),\\&(0,12,0),(12,12,0)];(0,0,9)(4,8,12)[(0,0,0),(12,0,0),(0,12,0),\\&(12,12,0)]\}\end{aligned} \tag{1}$$

where, taking into account indicated multiplication the coordinates of atoms can be found in unit of 1/24.

The above presentation makes use from the operator-like behaviour of relative coordinates. It can be seen due to the general convention

$$\exp\left[2\pi i(x_j h + y_j k + z_j l)\right] = (2\pi/s)\,(p_j, q_j, r_j)\,(hkl)\,, \tag{2}$$

where $2\pi/s$ is the common phase unit due to relative coordinates of atoms, (x_j, y_j, z_j), (p_j, q_j, r_j) are the particular coordinate operators and (hkl) plays the role of their eigenfunction.

From (2) immediately results

$$(p_i, q_i, r_i)\,(p_j, q_j, r_j) = [(p_i + p_j), (q_i + q_j), (r_i + r_j)] = (p_k, q_k, r_k)\,. \tag{3}$$

Moreover, we introduce spatial operator $\hat{G}_n$ being the sum of particular coordinate operators (p_j, q_j, r_j)

$$\hat{G}_n = (p_1, q_1, r_1) + (p_2, q_2, r_2) + \dots + (p_n, q_n, r_n)\,, \tag{4}$$

which is the basic one for the location all of atoms in the considered cell.

Similarly, can be defined atomic operator, $\hat{G}_a$

$$\hat{G}_a = f_M(p_i, q_i, r_i) + f_X(p_j, q_j, r_j) \tag{5}$$

where f_M and f_X are the atomic scattering factors.

Now, accordingly to the considered coordinates (1) we can define the quasistructure operator, $\hat{\Gamma}$

$$\hat{\Gamma} = f_M[\hat{G}_4 + (4, 8, 12)\,\hat{G}_4] + f_X[(0, 0, 9)\,\hat{G}_4 + (0, 0, 9)\,(4, 8, 12)\,\hat{G}_4]\,, \tag{6}$$

where

$$\hat{G}_4 = (0, 0, 0) + (12, 0, 0) + (0, 12, 0) + (12, 12, 0) \tag{7}$$

and all of location of $\hat{G}_4$ in equation (6) are quite independent. It shows the evolved shape of $\hat{\Gamma}$ then makes possible the study of its particular form if any vacancies are present in the cell. In other words, for SW case the structure amplitude is the eigenvalue of $\hat{\Gamma}$ so it can be written

$$\hat{\Gamma}(2\pi/s)\,(hkl) = F_{hkl}\,. \tag{8}$$

To confirm the above supposition, we notice the absence of one atom in the cell changes respectively one of four operators, $\hat{G}_4$ in equation (6) into other operator $\hat{G}_3$ which can take the form

$$\hat{G}_3 = (0, 0, 0) + (12, 0, 0) + (0, 12, 0)\,, \tag{9}$$

or each other which arises from $\hat{G}_4$ by removing of any (p, q, r). By virtue of symmetry of equation (6) and due to the above description of $\hat{G}_3$ it can be easy to seen that for

64 different choices of one VP in the cell, only four different types of $\hat{\Gamma}$ are obtained, each of these corresponding to 16 of the former, respectively. They are as follows

$$\hat{\Gamma}^{(1)} = f_M[\hat{G}_3 + (4, 8, 12)\,\hat{G}_4] + f_X[(0, 0, 9)\,\hat{G}_3 + (0, 0, 9)\,(4, 8, 12)\,\hat{G}_4]\,, \quad (10)$$

$$\hat{\Gamma}^{(2)} = f_M[\hat{G}_4 + (4, 8, 12)\,\hat{G}_3] + f_X[(0, 0, 9)\,\hat{G}_4 + (0, 0, 9)\,(4, 8, 12)\,\hat{G}_3]\,, \quad (11)$$

$$\hat{\Gamma}^{(3)} = f_M[\hat{G}_3 + (4, 8, 12)\,\hat{G}_4] + f_X[(0, 0, 9)\,\hat{G}_4 + (0, 0, 9)\,(4, 8, 12)\,\hat{G}_3]\,, \quad (12)$$

$$\hat{\Gamma}^{(4)} = f_M[\hat{G}_4 + (4, 8, 12)\,\hat{G}_3] + f_X[(0, 0, 9)\,\hat{G}_3 + (0, 0, 9)\,(4, 8, 12)\,\hat{G}_4]\,, \quad (13)$$

Now, they can be rewritten in compact form

$$\hat{\Gamma}^{(1)} = [\hat{G}_3 + (4, 8, 12)\,\hat{G}_4]\,[f_M + (0, 0, 9)\,f_X]\,, \quad (14)$$

$$\hat{\Gamma}^{(2)} = [\hat{G}_4 + (4, 8, 12)\,\hat{G}_3]\,[f_M + (0, 0, 9)\,f_X]\,, \quad (15)$$

$$\hat{\Gamma}^{(3)} = \hat{G}_3[f_M + (0, 0, 9)\,(4, 8, 12)\,f_X] + \hat{G}_4[(4, 8, 12)\,f_M + (0, 0, 9)\,f_X]\,, \quad (16)$$

$$\hat{\Gamma}^{(4)} = \hat{G}_4[f_M + (0, 0, 9)\,(4, 8, 12)\,f_X] + \hat{G}_3[(4, 8, 12)\,f_M + (0, 0, 9)\,f_X]\,. \quad (17)$$

Equations (14) and (15) include the well known atomic operator, $\hat{G}_a$ of IW and, as will be shown below, lead to just the same value of square of structure amplitude, $|F_{hkl}|^2$. The values of the latter for $\hat{\Gamma}^{(3)}$ and $\hat{\Gamma}^{(4)}$ are different, however, and more complicated.

3. Calculation of $|F_{hkl}|^2$ for SW case

Accordingly to equation (8) we have

$$\hat{\Gamma}^*(2\pi/s)\,(hkl)\,\hat{\Gamma}(2\pi/s)\,(hkl) = |F_{hkl}|^2\,, \quad (18)$$

where superscript denotes the conjugate value.

Notice, that $\hat{G}_4^* = \hat{G}_4$ and $\hat{G}_3^* = \hat{G}_3$. Moreover, their eigenvalues are respectively

$$\hat{G}_4(2\pi/s)\,(hkl) = \begin{cases} 4 & \text{for } h, k \text{ even} \\ 0 & \text{for other cases}\,, \end{cases} \quad (19)$$

$$\hat{G}_3(2\pi/s)\,(hkl) = \begin{cases} 3 & \text{for } h, k \text{ even} \\ -1 & \text{for other cases}\,. \end{cases} \quad (20)$$

Then, following $\hat{\Gamma}^{(1)}$ and $\hat{\Gamma}^{(2)}$ we obtain

$$|F_{hkl}^{(1),(2)}|^2 = \begin{cases} \{1 + 48\cos^2[(\pi/6)\,(h + 2k + 3l)]\}\,|Q_{hkl}^{(a)}|^2 & \text{for } h, k \text{ even} \\ |Q_{hkl}^{(a)}|^2 & \text{for other cases}\,, \end{cases} \quad (21)$$

or

$$|F_{hkl}^{(1),(2)}|^2 = |S_{hkl}^{(1),(2)}|^2\,|Q_{hkl}^{(a)}|^2\,, \quad (22)$$

where

$$|Q_{hkl}^{(a)}|^2 = f_M^2 + f_X^2 + 2f_M f_X \cos[(\pi/4)\,3l]\,, \quad (23)$$

is the atomic part of $|F_{hkl}|^2$ for IW, but for the latter case

$$|S_{hkl}|^2 = 4\cos^2[(\pi/12)\,(4h + 8k + 6l)]\,. \quad (24)$$

Using $\hat{\Gamma}^{(3)}$ as well $\hat{\Gamma}^{(4)}$ we obtain for h, k even

$$\begin{aligned} |F_{hkl}|^2 = \{1 + 48\cos^2[(\pi/6)\,(h + 2k + 3l)]\}\,\{f_M^2 + f_X^2 + \\ + 2f_M f_X\{24\cos[(\pi/4)\,3l] + A\cos[(\pi/12)\,(4h + 8k + 3l)] + \\ + B\cos[(\pi/12)\,(4h + 8k + 21l)]\}\}, \end{aligned} \quad (25)$$

where $A = 16$, $B = 9$ for $\hat{\Gamma}^{(3)}$ and vice versa for $\hat{\Gamma}^{(4)}$, respectively.

Significantly simpler formula holds for other cases

$$|F_{hkl}|^2 = f_M^2 + f_X^2 + 2 f_M f_X \cos\left[(\pi/12)(4h + 8k + rl)\right], \tag{26}$$

where r equals 21 for $\hat{\Gamma}^{(3)}$ or 3 for $\hat{\Gamma}^{(4)}$, respectively. Contrary to case of $\hat{\Gamma}^{(1)}$ and $\hat{\Gamma}^{(2)}$ it results from (22) to (26) that simple analogy with IW is here impossible. The equation (25) as well (26) can be, however, replaced by one expression

$$|F_{hkl}|^2 = |S_{hkl}^{(\mathrm{ef})}|^2 \, |Q_{hkl}^{(\mathrm{a})}|^2 , \tag{27}$$

where $|S_{hkl}^{(\mathrm{ef})}|^2$ is the effective value of purely geometrical part of square of structure amplitude.

Substituting (23) and (25) or (26) respectively into (27) we obtain numerical values of $|S_{hkl}^{(\mathrm{ef})}|^2$ which now can be simply compared with these of IW. The results are exemplified in Table 1 together with values obtained from (21). The column both for SW and IW is ascribed to their own (hkl), respectively.

Table 1
Exemplification of spatial part of square of structure amplitude, $|S_{hkl}^{(\mathrm{ef})}|^2$ for SW and IW, respectively. In the former case, the average of results for $\hat{\Gamma}^{(3)}$ and $\hat{\Gamma}^{(4)}$ types of location is shown*)

(hkl)	$\lvert S_{hkl}^{(\mathrm{ef})}\rvert^2$		
	$\hat{\Gamma}^{(1)}$ or $\hat{\Gamma}^{(2)}$	$\langle \hat{\Gamma}^{(3)}, \hat{\Gamma}^{(4)} \rangle$	IW
(10.0)	1	0.8	1
(00.2)	49	49	4
(10.1)	1	2.1	3
(10.2)	1	1	1
(11.0)	1	0.2	4
(10.3)	1	0.6	3
(20.0)	13	12.4	1
(11.2)	1	1	4
(20.1)	37	37.4	3
(00.4)	49	49	4
(20.2)	13	13	1
(10.4)	1	1.7	1
(20.3)	37	36.8	3
(21.0)	1	0.4	1
(21.1)	1	1.4	3
(11.4)	1	2.5	4
(10.5)	1	0.6	3
(21.2)	1	1	1

*) Calculations were made for $f_{\mathrm{Cd}} = 48$, $f_{\mathrm{S}} = 16$

4. Results and discussion

Our calculation takes into account 64 different choices of one VP in $(2a_0, 2a_0, 1c_0)$ region of wurtzite which appear due to neglecting of any correlation between coordinates of metal and non-metal vacancy. As result, only three different characteristics of $|F_{hkl}|^2$ are obtained. Besides, the average of numerical values of $|S_{hkl}^{(\mathrm{ef})}|^2$ for $\hat{\Gamma}^{(3)}$ and $\hat{\Gamma}^{(4)}$, respectively, is nearly the same as the value which corresponds to $\hat{\Gamma}^{(1)}$ or $\hat{\Gamma}^{(2)}$. This can be the basis for the concept of SW. Indeed, it results from Table 1 that SW can be approximately presented by one characteristics of $|F_{hkl}|^2$, only.

Experimental evidence of SW is, however, not so strong. First of all, it is a superstructure then forbidden reflections of IW such as (111), (301) . . . could be observed as (221), (601) . . . of SW, respectively, but they are unattainable (HORODECKI et al.). Nevertheless, the reflections of SW due to h or k or both h and k odd would be, in accordance with Table 1, very weak and in fact they are also unattainable. All of remain reflections of SW i.e. for h, k even have their $|F_{hkl}|^2$ greater more than order of magnitude in comparison with IW. They are superimposed, respectively, on $((1/2)\,h,\,(1/2)\,k,\,l)$ reflections of IW then would lead to increase of resultant intensity as well to significant broadening of them due to well known property of superstructure.

Turning out to phase II (HORODECKI) as the most representative for SW concept there are predominant always $(10.l)$ reflections with l ranging from 2 to 6. With regarding to experimental value of $|F_{hkl}|^2$ they agree with the above supposition. Besides, the less l, the greater is broadening so, consequently the vanishing of (10.1) as well (10.0) is not surprised. On the contrary, the greater l, the nearer is plane to (00.l) which is, as a rule, only reasonably broadened. It can be interpreted in terms of stacking faults (EDWARDS, LIPSON; WILSON). On the other hand, the reflections of phase M not reveal of any distincted broadening then its anomalously high (10.3) intensity seems to be the result of VP presence, only. In general, however, both SW and stacking faults would be taken into consideration as joint effect of vapour growth of pyroelectric thin film. The polytype structure can not be excluded also, as the extraordinary reflections of phase II (HORODECKI et al.) not belong to IW nor to SW lattice. Besides, the effect of strong anisotropy of crystallite dimensions would be recognized.

Apart from the above discussed regularities, notice the intensities of reflections of phase I being considerably weaker and of powder type. This eliminates rather SW characteristics.

5. Conclusions

The presented SW concept seems to be the way for analysis of real structure of crystal containing point defects. To generalize it, however, further investigations are need although the simple extension is always possible. Nevertheless, the operator processing used here may be developped for a new description of other structural questions e.g. of polytype structure. The presented results explain partly the peculiarities of X-ray pattern of CdS thin films (HORODECKI et al.; HORODECKI). The interpretation of extraordinary reflections will be given in the next paper.

References

EDWARDS, O. S., LIPSON, H.: Proc. Roy. Soc. London **180A**, 268 (1942)
HORODECKI, A. J., ŁEPEK, M., PRECHT, W.: Kristall und Technik **13**, K18 (1978)
HORODECKI, A. J.: Cryst. Res. Technol. (in press)
SCHMIDT, W., PAHLE, T.: Kristall und Technik **12**, K39 (1977)
WILSON, A. J. C.: Proc. Roy. Soc. London **180A**, 277 (1942)

(Received May 18, 1987)

Author's address:
Institute of Experimental Physics, University of Gdańsk
ul. Wita Stwosza 57
80-952 Gdańsk
Poland

Book Review

S. E. Trullinger, V. E. Zakharov, V. L. Pokrovsky (eds)

Solitons

Modern Problems in Condensed Matter Sciences. Vol. 17

North-Holland Physics Publishing, Amsterdam-Oxford-New York-Tokyo 1986. 899 p. Dfl. 400,00, ISBN 0-444-87002-4

The permanent improvement of the experimental technique gives us many hints that a theoretical description of certain phenomena is no longer adequate by means of linear methods so that non-linear models have to be used. This is a challenge to study more deeply nonlinear ordinary and partial differential equations, leading to several surprises. One was surely the rediscovery of the "solitary waves", localized stable excitations propagating with constant velocity in an infinite system. Though they were already discovered in the last century the term "soliton" was coined only some twenty years ago. Nowadays solitons play a fundamental role in different fields ranging from condensed matter to high energy physics and cosmology. If a pure elastic scattering of solitary waves occurs then these excitations are called solitons and the systems are "integrable" ones. Solitons are now special solutions of integrable nonlinear evolution and wave equations in $(1 + 1)$-space-time dimensions and may be bell or kink shaped depending whether the asymptotic values are equal or different. Several attempts are made to extend the notion of a soliton and the methods of solution to higher space dimensions, e.g. via the Riemann problem. But surprisingly too even the extension of the equations itself is not straightforward as is seen in the case from the Korteweg-de Vries to the KP-equation.

In this extensive exposition about solitons an interesting survey is given about their importance in condensed matter problems. Starting with a soliton mechanism of energy transfer along biological polymers, thermodynamics of a two-dimensional soliton system is developed to treat surface phenomena. The applications to spin chains and waves, to lattice dynamical models, to the propagation in the Josephson transmission line and the occurence of exotic textures in superfluid ^{3}He are treated in detail. Special considerations are devoted to the role of solitons in charge and spin density waves in electron-phonon systems. Instructive fundamental theoretical ideas are incorporated in the chapters about soliton stability, integrable magnetic models and the thermodynamical and quantum-mechanical aspects of solitons on a Toda lattice. Most of the contributions are self-contained, providing the reader with a valuable information.

This volume will be of interest to a broad scientific audience looking for an accessible survey to this rapidly growing subject. P. Möbius

Cryst. Res. Technol.	22	1987	12	1477–1484

M. SZURGOT, K. SANGWAL

Technical University of Łódź, Poland

Influence of Free Convection on the Formation of Growth Defects in Potassium Bichromate Crystals Grown from Aqueous Solutions[1])

The influence of free convection on the formation of inclusions, growth bands and dislocations in potassium bichromate crystals grown by self-nucleation and on freely hanging seeds from aqueous solutions in a wide range of temperature is studied. It is found that in free convection all the three types of defects appear mainly in the top sectors but in stirred solutions the slow-growing faces are more defective.

1. Introduction

Effect of convection on crystal growth embraces a wide range of problems. Convection affects the rates and kinetics of growth of faces, surface micromorphology and compositional and impurity striations (WILCOX; CARRUTHERS; SIMON (a, b); SIMON, PANTALONI; GITS-LEON et al.; VADILO; PISMENNYI (a, b); CHEN et al.; KOLASIŃSKI; PETROVSKY; NERAD, SHLICHTA). In recent years there is an increasing interest in the investigation of the effect of convection during the growth of crystals from solution mainly because convection phenomenon may be directly observed not only by interferometric (BEDARIDA et al.) or schlieren techniques (CHEN et al.) but the changes in growth kinetics, surface micromorphology and defect distribution may also be correlated with it (SIMON, PANTALONI; GITS-LEON et al.; VADILO; PISMENNYI (a, b); KOLASIŃSKI; PETROVSKY; SIMON (a, b); NERAD, SHLICHTA). The results reported in the literature show that with free convection due to gravity inclusions form in the top of the side faces and on the top face (WILCOX; SIMON, PANTALONI; VADILO), whereas forced convection due to mechanical stirring and pumping of the solution introduces inclusions on faces located in the wake where recirculating eddies take place (GITS-LEON et al.; PISMENNYI (b); JANSSEN-VAN ROSMALEN, BENNEMA; VAN ENCKEVORT et al.). In forced convection, the capture of inclusions increases as one moves away from the seed in the direction of solution flow, but details of the effects of free convection on inclusion formation are relatively unknown.

The aim of this work is to investigate the influence of free convection on the formation of inclusions, growth bands and dislocations in potassium bichromate (KBC) crystals grown from aqueous solutions in a wide range of temperature.

2. Experimental

Crystal growth was carried out in 2-litre round-bottomed flasks on seeds with or without stirring, or in 2-litre beakers by self-nucleation without stirring. The experimental details

[1]) Paper based on poster PS 1/317 presented at the 8th Intern. Conf. on Crystal Growth, York 1986.

for crystal growth are described in earlier papers (SZURGOT et al.; SANGWAL, SZURGOT (a); SZURGOT, SANGWAL). To investigate the distribution of inclusions and the planes in which they lie the crystals were examined, as in a previous paper (SZURGOT et al.), in transmission using illumination perpendicular as well as parallel to different planes of observation. Dislocations and growth bands were revealed by etching the as-grown and cleaved surfaces at room temperature in HCOOH for 1 – 3 seconds (SZURGOT et al.; SANGWAL, SZURGOT (b)). The etched surfaces were examined under an optical microscope in reflection.

3. Observations

3.1. Distribution of inclusions

It was found that in KBC crystals grown on seeds from stirred solutions at growth temperatures below ca. 40 °C the sectors containing the maximum inclusions belong to the slowest-growing (100) and (00$\bar{1}$) faces, but at elevated temperatures the corner (11k) and (1$\bar{1}k$) faces (with $k = 0, \pm 1$) contain a high inclusion content. The {012}, {011}, {101} and {$\bar{1}$01} sectors contain relatively less inclusions. Some of the sectors illustrating the abovementioned characteristic features may be observed from the

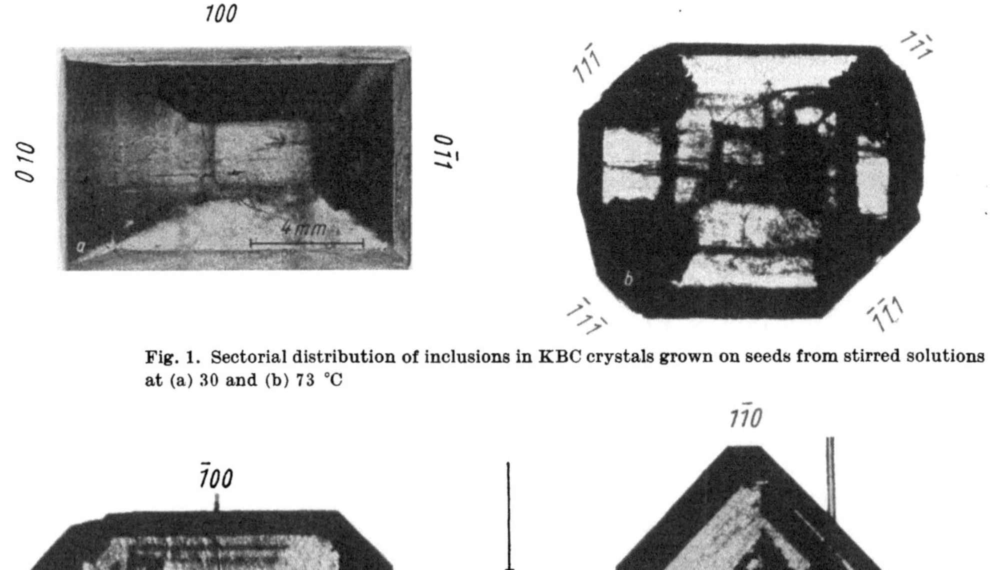

Fig. 1. Sectorial distribution of inclusions in KBC crystals grown on seeds from stirred solutions at (a) 30 and (b) 73 °C

Fig. 2. Distribution of inclusions in KBC crystals grown on freely hanging seeds from unstirred solutions at different rates of cooling: (a) 0.27 °C/h in the temperature interval 52 to 26 °C and (b) 0.60 °C/h in the temperature interval 42.6 – 21.5 °C. The arrow indicates the direction of gravitational field

crystals shown in Figure 1. One also notes that the distribution of inclusions in parallel opposite faces is different. This difference is particularly marked for the {001} and {100} faces, but decreases with an increase in growth temperature. In the absence of stirring, the natural anisotropy in defect distribution disappears at the expense of convection-induced defects which mainly appear in the topmost sectors, as shown in Figures 2 and 3.

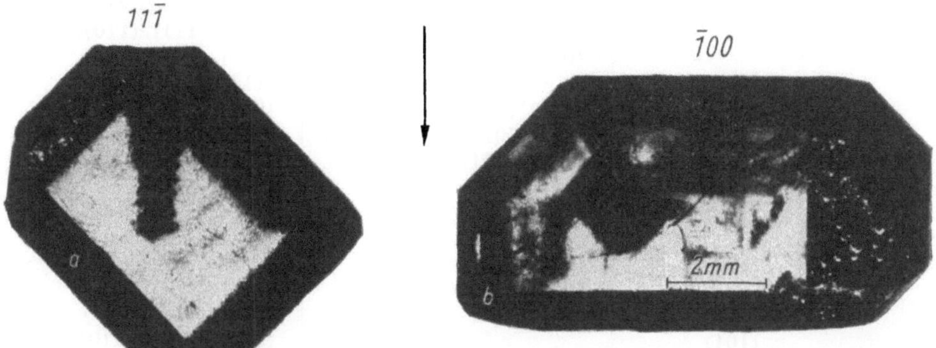

Fig. 3. Inclusion distribution in KBC crystals grown by self-nucleation at the bottom of growth vessel from unstirred solutions at 97 °C. The planes in contact with the bottom are: (a) $(\bar{1}\bar{1}1)$ and (b) (100). The arrow indicates the direction of gravitational field

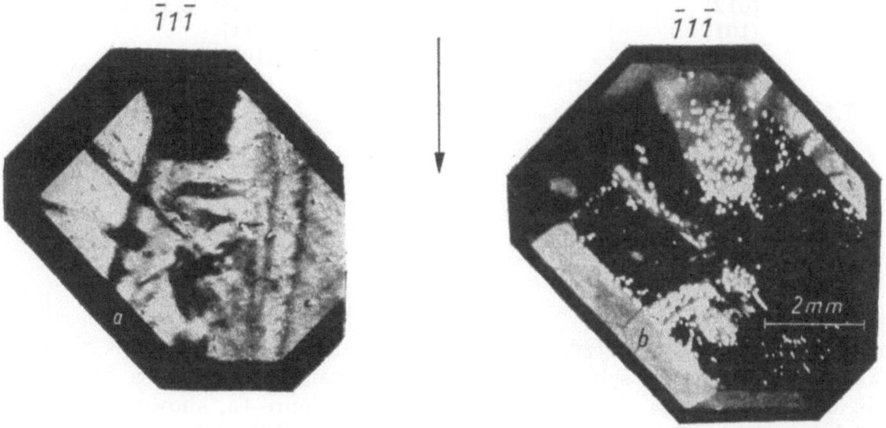

Fig. 4. Distribution of (a) inclusions and (b) dislocations in a KBC crystal grown by self-nucleation from unstirred solution in the temperature interval 60 – 53 °C at cooling rate 1.90 °C/h. The plane of observation is $(00\bar{1})$. The arrow indicates the direction of gravitational field. (b) shows the crystal after etching where the white dots are dislocation etch pits. A high density of dislocations may be noted in the $(\bar{1}1\bar{1})$ growth sector as well as in other regions of high content of inclusions seen as dark areas in (a)

Figures 3 and 4 illustrate some examples of the distribution of inclusions whereas Table 1 presents the data on the presence of inclusions and their relative intensity in crystals grown by self-nucleation at the bottoms and the walls of growth vessels. In the table sequence of sectors is arranged according to a decrease in the intensity of capture of inclusions in a crystal. One notes from Figures 2–4 and from Table 1 that, in general irrespective of the crystal orientation, inclusions occur predominantly in the vertical growth sectors opposite to the face in contact with the bottom of a growth vessel.

Table 1
Sectors of inclusions in KBC crystals grown by self-nucleation at 11 and 97 °C

Plane of contact with growth vessel	sectors of inclusions	
	growth temperature 11 °C	growth temperature 97 °C
(001)	$(00\bar{1})^{DV}$*), (100), $(0\bar{1}0)$, $(0\bar{1}1)$, $(01\bar{1})$, (010), (001)	$(00\bar{1})^{V}$, $(\bar{1}\bar{1}1)$, $(11\bar{1})$, $(\bar{1}\bar{1}\bar{1})$, (111), $(\bar{1}10)$, $(1\bar{1}0)$, $(\bar{1}11)$, $(1\bar{1}1)$, $(1\bar{1}\bar{1})$, (001)
$(00\bar{1})$	$(001)^{V}$, $(100)^{D}$, $(0\bar{1}1)^{D}$, $(0\bar{1}0)$, (010), (012), $(\bar{1}01)$, $(00\bar{1})$	$(001)^{DV}$, $(\bar{1}\bar{1}1)$, $(11\bar{1})$, (111), $(\bar{1}10)$, $(\bar{1}1\bar{1})$, $(1\bar{1}1)$, $(\bar{1}11)$, $(0\bar{1}1)$, $(0\bar{1}0)$, $(00\bar{1})$
(100)	$(00\bar{1})^{D}$, $(0\bar{1}1)^{D}$, $(0\bar{1}0)$, (010), (100), $(\bar{1}00)$	$(\bar{1}00)^{DV}$, $(\bar{1}\bar{1}1)$, $(\bar{1}10)$, $(0\bar{1}\bar{1})$
$(\bar{1}00)$	$(100)^{DV}$, $(0\bar{1}1)$, (010), $(00\bar{1})$	–
$(\bar{1}01)$	–	$(1\bar{1}0)$, $(1\bar{1}\bar{1})$
$(10\bar{1})$	–	$(\bar{1}01)^{V}$, $(\bar{1}\bar{1}0)^{D}$, $(0\bar{1}1)$
(010)	$(0\bar{1}0)^{DV}$, $(0\bar{1}1)^{D}$, $(00\bar{1})$, (100), (012)	–
$(0\bar{1}0)$	$(010)^{DV}$, (100), (012), $(00\bar{1})$, $(0\bar{1}1)$	–
(012)	–	$(0\bar{1}\bar{2})^{DV}$, $(1\bar{1}1)$, $(\bar{1}\bar{1}1)$
(101)	–	$(\bar{1}11)$
$(\bar{1}0\bar{1})$	–	$(101)^{DV}$
$(\bar{1}11)$	–	$(1\bar{1}\bar{1})^{DV}$
$(\bar{1}\bar{1}1)$	–	$(11\bar{1})^{DV}$, $(10\bar{1})$
(110)	–	$(\bar{1}\bar{1}0)^{V}$, $(1\bar{1}1)$
$(\bar{1}1\bar{1})$	–	$(101)^{D}$, $(1\bar{1}0)$, $(\bar{1}10)$, $(1\bar{1}\bar{1})$
$(1\bar{1}1)$	–	$(\bar{1}1\bar{1})^{DV}$

*) D – dominant, V – vertical

3.2. Distribution of growth bands and dislocations

The typical examples of the distribution of growth bands and dislocations observed on KBC crystals grown by self-nucleation are represented in Figures 4b and 5. Figure 4b, which is an etch pattern of the crystal of Figure 4a, shows that a high density of dislocation etch pits appears in the $(\bar{1}1\bar{1})$ sector. Figure 5a shows that the dislocation density is much higher in the (100) growth sector than in the $(0\bar{1}0)$ and $(00\bar{1})$ sectors. Figure 5b illustrates that intense growth bands appear in the $(00\bar{1})$, (100), $(10\bar{1})$ and (001) sectors, whereas Figure 5c shows that they are better produced in the $(00\bar{1})$ sector.

In crystals grown from stirred solutions on seeds, the (100) and $(00\bar{1})$ growth sectors contain a higher inclusion content than other growth sectors, while in crystals grown from unstirred solutions on seeds and by self-nucleation, inclusions are predominantly captured in the top sectors. From these observations as well as from Figures 4b and 5 it may be concluded that, in each case irrespective of the method of growth, the density of dislocations and the intensity of growth bands is high in those sectors which have a high content of inclusions. A similar correlation between dislocation density, and intensity of growth bands and inclusions was found previously (SZURGOT et al.).

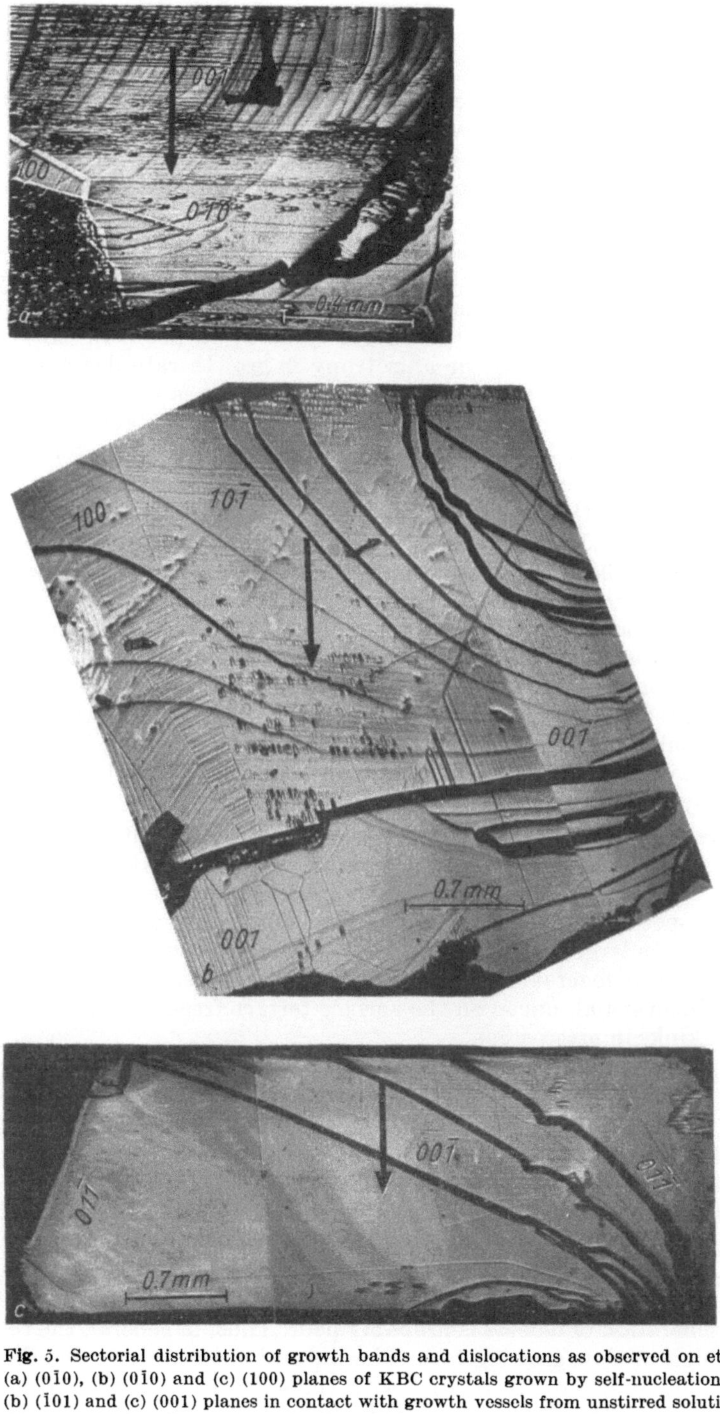

Fig. 5. Sectorial distribution of growth bands and dislocations as observed on etched cleaved (a) (0$\bar{1}$0), (b) (0$\bar{1}$0) and (c) (100) planes of KBC crystals grown by self-nucleation on (a) (001), (b) ($\bar{1}$01) and (c) (001) planes in contact with growth vessels from unstirred solutions at (a) 63, (b) 97 and (c) 90 °C, respectively. The arrow indicates the direction of gravitational field. A high density of dislocations and growth bands may be observed in the top sectors of these crystals

4. Discussion

The density of the generated defects is directly related with kinetic and environmental factors. The kinetic factors are associated with the growth rate anisotropy at the given conditions of supersaturation and temperature, whereas the environmental factors include the characteristics of solution flow. Convection effects fall in the latter category.

In the crystals growing on seeds and by self-nucleation from unstirred solutions both kinetic and convection effects are present. At low temperatures kinetic effect is noticeable, as may be inferred from the distribution of defects predominantly present in the slow-growing sectors in crystals grown at 11 °C (Tab. 1). However with an increase in temperature when the anisotropy in growth rate decreases (RODRIGUEZ et al.; SANGWAL, SZURGOT (a)), convection effect becomes very pronounced, as may be seen from a good correlation between the planes in contact with the bottom of growth vessels and the planes containing the highest inclusion content.

The interrelation between inclusions, growth bands and dislocation content in KBC crystals means that the causes of their generation are the same. Fluctuations in growth conditions, high supersaturations and impurities present in solutions are the main causes. These parameters lead to the formation of bunches (RODRIGUEZ et al.; SANGWAL et al.; SANGWAL) through which the defects are generated. The bunched layers transform into macosteps which rearrange into equilibrium facets belonging to the crystallographic zone of faces of the crystal normal to the surface under observation (SANGWAL et al.). The macrokinks or teeth formed by the adjoining equilibrium facets act as sites of the capture of liquid inclusions and impurity aggregates, which subsequently lead to the generation of dislocations. Since the process of bunching is favorable when the interlayer distance is smaller, it is expected that the defect density should be higher in faser growing sectors.

The above inference of a direct dependence of defect density on growth rate is in contradiction with the present experimental results. This discrepancy may be attributed to the participation in growth of chemical entities much larger than ionic or molecular species, but these larger chemical entities need not be the actual active growth units. The larger chemical units not only enter into monomolecular kink sites in the ledges or into macrokinks in the macrosteps, but also act as two-dimensional, or even three-dimensional, nuclei on the surface terrace free from growth layers as well as on macrokinks in macrosteps.

According to the theory (OHARA, REID), the rates of nucleation and incorporation of growth units increase with supersaturation and temperature, but the effects of facilitated nucleation and growth unit incorporation are opposite. The former effect is responsible for the generation of defects, particularly in slow-growing faces where free growth sites are relatively few. The latter effect leads to the suppression of generation of defects due to an increase in the diffusion distance for growth units with an increase in temperature and due to an increase in the number of growth sites with supersaturation. Since the relative concentration of larger chemical entities is expected to increase with supersaturation, the effect of supersaturation on the suppresion of generation of defects is probably poor. Thus, in general, the relative defect density is high in slow-growing face. The natural anisotropy in the distribution of defects is, therefore, directly related with the anisotropy in growth rates i.e. with growth kinetics. With a decrease in growth anisotropy, the anisotropy in the distribution of defects also decreases. Moreover, the generation of defects by the two- or three-dimensional nucleation mechanism may be accentuated by trace amounts of insoluble decomposition products (e.g. Cr_2O_3) of KBC via increased bunching of growth layers by, and induced nucleation at, them.

The preceding ideas may also be used to explain the effect of free convection on the generation of defects on top faces. The top faces of crystals growing in free convection conditions lie at the tail of rising convection currents. Consequently, an instability develops in the flux of solution across the crystal due to eddies or solution starvation in the regions of these faces (CHEN et al.; BEDARIDA et al.). These instabilities accelerate the process of bunching of growth layers, which lead to the generation of defects on macrokinks in the macrosteps. Furthermore, tiny crystallites developed in the bulk solution and subsequently falling on the top faces, and two- or three-dimensional nuclei developed on the surface terrace free from growth layers may also introduce defects. The effects of free convection on the introduction of defects in KBC are particularly pronounced when they combine with the natural anisotropy in defect distribution.

The authors express their gratitude to Doc. dr J. Karniewicz for his interest in the work. This work was carried out under Research Project CPBP 01.20. 3—2.4.

References

BEDARIDA, F., BOCCACCI, P., ZEFIRO, L., PONTIGGIA, C.: Physico-Chem. Hydrodyn. 2, 327 (1981)

CARRUTHERS, J. R.: in: Preparation and Properties of Solid State Materials, Vol. 3, Eds W. R. WILCOX and R. A. LEFEVER, New York 1977

CHEN, P. S., SHLICHTA, P. J., WILCOX, W. R., LEFEVER, R. A.: J. Cryst. Growth 47, 43 (1979)

ENCKEVORT, VAN W. J. P., JANSSEN-VAN ROSMALEN, R., KLAPPER, H., VAN DER LINDEN, W. H.: J. Cryst. Growth 60, 67 (1982)

GITS-LEON, S., LEFAUCHEUX, F., ROBERT, M. C.: J. Cryst. Growth 44, 345 (1978)

JANSSEN-VAN ROSMALEN, R., BENNEMA, P.: J. Cryst. Growth 42, 224 (1977)

KOLASIŃSKI, W.: Mater. Res. Bull. 19, 867, 873 (1984)

NERAD, B. A., SHLICHTA, P. J.: J. Cryst. Growth 75, 591 (1986)

OHARA, M., REID, R. C.: Modeling Crystal Growth Rates from Solution, New Jersey 1973

PETROVSKY, V. A.: Rost Kristallov v Heterogennykh Rastvorakh (Crystal Growth in Heterogeneous Solutions), Leningrad 1983

PISMENNYI, V. A.: Zap. Vsekhsoy. Miner. Obshch. 89, 699 (1960) (a)

PISMENNYI, V. A.: Miner. Sborn. Lvov. Gos. Univ. 24, 37 (1970) (b)

RODRIGUEZ, R., AGUILO, M., TEJADA, J.: J. Cryst. Growth 47, 518 (1979)

SANGWAL, K.: Etching of Crystals-Theory, Experiment and Application, Amsterdam 1987

SANGWAL, K., RODRIGUEZ-CLEMENTE, R., VEINTEMILLAS-VERDAGUER, S.: J. Cryst. Growth 78, 144 (1986)

SANGWAL, K., SZURGOT, M.: J. Cryst. Growth 80, 351 (1987) (a)

SANGWAL, K., SZURGOT, M.: Cryst. Res. Technol. 17, 49 (1982) (b)

SIMON, B.: J. Cryst. Growth 43, 640 (1978) (a)

SIMON, B.: J. Cryst. Growth 61, 167 (1983) (b)

SIMON, B., PANTALONI, J.: Physico-Chem. Hydrodyn. 5, 19 (1984)

SZURGOT, M., SANGWAL, K., KARNIEWICZ, J.: Cryst. Res. Technol. 20, 645 (1985)

Szurgot, M., Sangwal, K.: J. Cryst. Growth **79**, 829 (1986)
Vadilo, P. S.: Zh. Eksp. Teor. Fiz. **8**, 1218, 1373 (1938)
Wilcox, W. R.: J. Cryst. Growth **65**, 133 (1983)

(Received May 20, 1987)

Authors' addresses:

Dr. M. Szurgot, Doz. Dr. hab. K. Sangwal*
Institute of Physics, Technical University of Łódź
Wólczańska 219
93005 Łódź, Poland

Permanent address:
Department of Physical and Solid State Chemistry
Pedagogical University of Częstochowa
Al. Zawadzkiego 13/15
42200 Częstochowa, Poland

Cryst. Res. Technol. | 22 | 1987 | 12 | 1485–1491

I. IORDANOVA, V. BOJINOV, L. ILIEV

Sofia University, Faculty of Physics, and
Faculty of Chemistry

Influence of Impurities on Texture and Structure Formation through Recrystallization of Aluminium Killed Cold Rolled Steel

The microhardness, electric resistance, volume of the unit cell, texture, form and size of ferrite grains and cementite particles depending upon recrystallization of cold rolled steel 08 killed with Al are investigated. Some conclusions about the morphology of solution of carbides and nitrides and about the influence of impurities on the recrystallization texture are drawn. Some recommendations about technology of annealing of preliminary aged cold rolled steel 08 killed with Al are given.

Комплексно были изучены микротвердость, электросопротивление, обьем элементарной ячейки кристаллической решетки, текстура, форма и размер ферритных зерен и цементитных частиц при отжиге холоднокатаной стали 08Ю. Сделаны заключения о морфологии растворения карбидов и нитридов и о влиянии примесных атомов на текстуру рекристаллизации. Даны рекомендации относительно технологии отжига предварительно состаренной холоднокатанной стали 08Ю.

1. Introduction

It is known that the conditions of recrystallization annealing (heating rate, temperature, time of keeping) of the deformed low carbon steels have an influence upon the material structure parameters. In some of the investigations (DILLAMORE et al.; DIAKONOVA et al.; MEYZAND et al.; PARNIERE) the kinetics of the recrystallization processes following high degrees of reduction is connected with the changes of the concentration mainly of C and N atoms in the solid solution on the α-Fe base. C and N atoms can be uniformly dissolved in the lattice, can be stored as clusters on dislocations or can form inclusions.

At present the prevalent is the opinion (DILLAMORE et al.) that the precipitation of fine-scale inclusions of AlN immediately before or at the very beginning of recrystallisation improve the structure and operation characteristics of the aluminium killed steel, used for stamping. In this connection in production practice it has been apprehended to get the rolling of the hot rolled strip at temperatures 550–600 °C. This provides fixing of carbon and nitrogen atoms in an oversaturated solid solution in α-Fe lattice and their onward precipitation as nitrides and carbides during annealing.

In the low carbon cold rolled steels for deep drawing it is necessary to have anisotropy of plastic properties along the normal to the sheet's surface. This anisotropy is estimated by $\bar{R}$-factor (LANCFORD et al.). Satisfactory values of $\bar{R}$-factor are received when the relation between texture components $\{111\}$ and $\{100\}$ is high.

The influence of AlN fine-scale precipitation on the recrystallization texture has been studied by many authors, and it has been published in many papers and reviews.

Unfortunately mainly because of experimental difficulty, there is not a clear-cut concept about the morphology and quantitative characteristics of this process yet.

The reverse situation about the influence of dissolving C and N atoms in α-Fe lattice during annealing upon the recrystallization texture has been slightly reflected in scientific literature. The experimental data about such a case are limited enough. In practice such a situation relates to a recrystallization annealing of a cold rolled strip which has been produced from a hot rolled strip with high rolling temperature or to an ageing of the cold rolled strip before the recrystallization annealing (with keeping at room temperature, for example). In this case, contrary to the above mentioned, C and N atoms are not in oversaturated solution before the annealing. They have been precipitated as nitrides or carbides which get solved during the annealing and increase C and N concentration in the solid α-Fe solution.

The experimental investigation of α-Fe solid solution's concentration and of the impurities forming clusters or inclusions is connected with great difficulty. It requires a special apparatus with both high sensibility and separating abilities.

The aim of this paper is to investigate the behaviour of C and N atoms during recrystallization by using more universal methods for estimating the physical parameters, which are sensitive to the concentration and distribution of C and N atoms. The second aim is to observe the influence of the concentration, increasing during annealing, of these atoms in α-Fe solid solution upon the recrystallization texture and structure formation.

2. Material and experiment

A cold rolled sheet of aluminium killed low carbon commercial steel with total reductions 56% and 69% and thickness of 1.5 and 1.0 mm, respectively, has been investigated. The chemical composition is shown in Table 1. The following physical parameters have been observed: the Vickers microhardness (with a load of 20 g and statistics from a hundred measurements for every specimen); texture by the Harris method (molybdenum X-ray

Table 1
Chemical composition, weight %

C	Mn	Si	S	P	Ni	Cr	Al	N
0.06	0.28	0.018	0.026	0.016	0.034	0.02	0.036	0.0066

diffraction and monochromatization of the diffracted beam); the size of ferrite grains and cementite inclusions (by metallographic investigations of polished specimens treated with 4% alcohol solution of HNO_3 and by the methods of stereographic metallography for a quantitative estimation of volume content of Fe_3C (SALTICOV); the specific electro-resistance (at a temperature of liquid nitrogen) by compensation method on electric sparkcut specimens with 0.1 m length and 0.001 m width; the exact sizes of the specimens and the distance between potential contacts were measured by a comparator with an accuracy of $\pm 10^{-6}$ m; the results in the series of measurements made for each specimen do not differ more than 0.3%); the starting temperature of recrystallization (by X-ray diffractometry and by metallography); volume of lattice unit (by X-ray diffraction of CuKα monochromatic rays and determination of the centre of weight in {211} diffraction profile; the averaged volumes are taken from the data of 10 measurements.

After the experimental specimens had been cut they were kept at a room temperature for about two months and then heated in an inert atmosphere with a heating rate of about 100 °C/h to temperatures: 100 °C, 200 °C, 300 °C, 400 °C, 450 °C, 500 °C, 600 °C, and 700 °C. After isothermal keeping for an hour the specimens were quenched in cold water and then the above mentioned properties were measured. The scecimens for the electric resistance measurements were annealed in argon atmosphere and the specimens for the other measurements in nitrogen. All measurements (except these for resistance) were made on the middle thickness sections of the specimens. It was done in order to prevent the influence of the non-uniform chemical composition along the thickness on the structure parameters. Isothermal annealing with different keeping times were made also. There were three annealing regimes of the specimens for texture measurements: regime A — heating to 500 °C with a rate 100 °C/h, keeping at this temperature for 2 hours, heating to 700 °C with the same rate, keeping at 700 °C for 1 hour, quenching; regime B — as regime A, but the intermediate keeping is at 550 °C instead of 500 °C; regime C — heating to 700 °C with a rate of 100 °C/h, keeping at this temperature for 1 hour, quenching.

3. Results and discussion

Changes of microhardness during annealing are shown on Figure 1. The dependence to 500 °C is complicated, as at first microhardness increases slightly, then it decreases and finally increases again. The curve for the material deformed with $\varepsilon_\Sigma = 69\%$ is displaced with about 100 °C to the lower temperatures in comparison to the one according to the material deformed with $\varepsilon_\Sigma = 56\%$. After 500 °C the Vickers hardness decreases gradually. The average values of microhardness and their deviations are shown in Table 2.

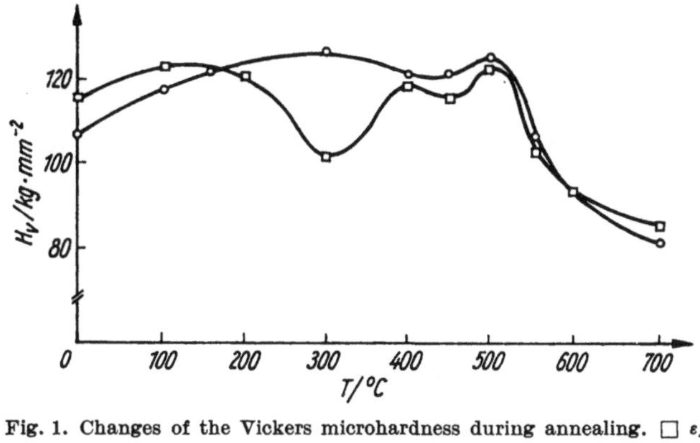

Fig. 1. Changes of the Vickers microhardness during annealing. □ $\varepsilon_\Sigma = 69\%$, ○ $\varepsilon_\Sigma = 56\%$

Microstructures of deformed and annealed at 500 °C specimens are shown on Figure 2. It is obvious that initial recrystallization begins at this temperature. This result is confirmed by X-ray diffraction. The recrystallization stage at this temperature is estimated quantitatively by metallographic method (SALTICOV) and is equal to 50%.

The dependence of specific electric resistance on the annealing temperature is shown on Figure 3. The values of the resistance do not change during annealing to the temperature of primary recrystallization, but then increase sensibly with increasing of the temperature to 700 °C.

The change of the resistance values during isothermal keeping at 700 °C is shown on Figure 4. It is obvious that the sensitive decrease of resistance is observed after the first hour of keeping. It remains almost the same when the keeping time increases and its values are only slightly smaller than observed in initially deformed material (see Fig. 3).

Table 2

No.	ε_Σ, %	temperature of annealing, °C	$\bar{H}_V$, kg/mm²	$\pm\sigma_{\bar{H}V}$, kg/mm²
1	56	cold rolling	107.1	0.9
2	56	100	118.4	1.2
3	56	200	123.2	1.0
4	56	300	128.6	0.9
5	56	400	122.1	1.0
6	56	450	122.1	0.9
7	56	500	125.3	0.8
8	56	550	94.2	0.6
9	56	600	97.0	1.2
10	56	700	87.1	0.5
11	69	cold rolling	114.9	1.0
12	69	100	121.5	0.6
13	69	200	121.9	1.0
14	69	300	108.3	0.6
15	69	400	118.6	0.1
16	69	450	118.1	0.9
17	69	500	124.0	0.6
18	69	550	107.3	0.8
19	69	600	97.3	1.3
20	69	700	87.6	0.3

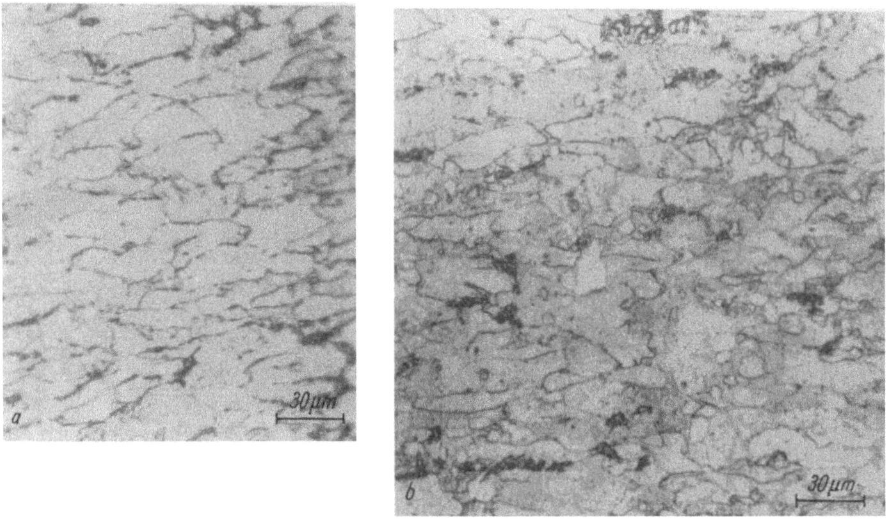

Fig. 2. Microstructure of deformed (a) and annealed at 500 °C (b) steel

The dependence of the relative part of cementite (Fe_3C) inclusions on the annealing temperature is shown on Figure 5. Because inclusions becoming very disperse at 700 °C it was impossible to observe them at this temperature. That is why this dependence is drawn only to 600 °C.

Table 3 shows the changes of the lattice volume unit of α-Fe during annealing.

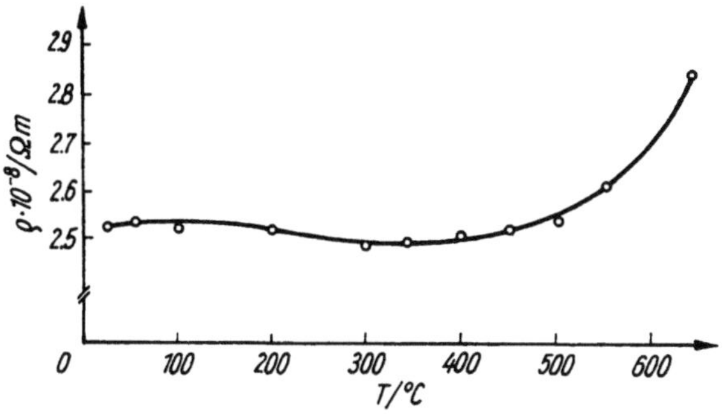

Fig. 3. Dependence of specific electric resistance on the annealing temperature

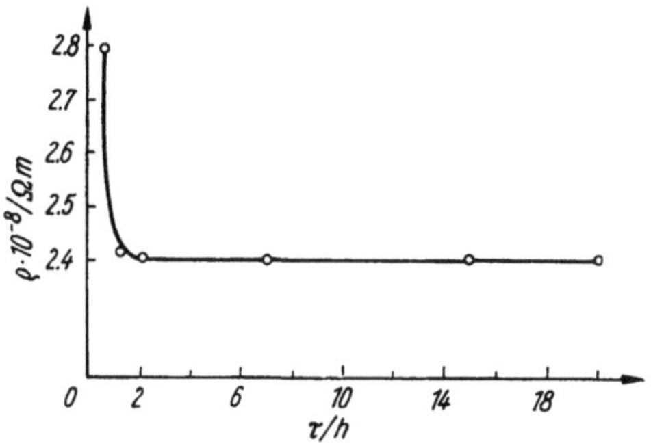

Fig. 4. Change of the resistance values during isothermal keeping at 700 °C

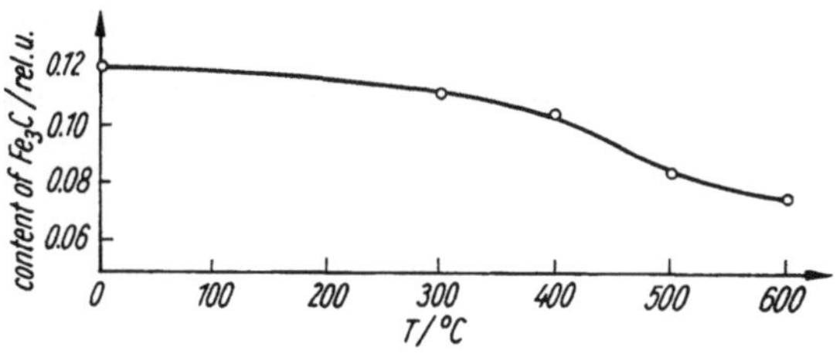

Fig. 5. Dependence of the relative part of cementite inclusions on the annealing temperature

Table 3

No.	temperature of annealing, °C	averaged lattice unit's volume, Å^3	$\pm\sigma\bar{v}$, Å^3
1	cold rolling	23.4233	0.0002
2	200	23.4217	0.0004
3	400	23.4267	0.0004
4	500	23.4253	0.0003
5	550	23.4273	0.0004

Generalization of the experimental results shows that at the temperature of primary recrystallisation (500 °C) two processes occur simultaneously in the investigated specimens. They have a contrary influence upon observed parameters. These processes are: recovery (connected with redistribution and partly anihilation of vacancies and dislocations) and increase of the concentration of C and N atoms in α-Fe lattice. Predominance of every one of these processes at a determined temperature interval, accordingly leads to decrease or increase of H_v or of the unit cell volume. In correspondence with the data shown on Figure 1, the processes begin earlier in the heavier deformed steel, which is thermodynamically less stable. The resistance values are independent on temperature to the beginning of recrystallization. It may be connected with the fact that electric resistance is sensitive mainly to the concentration of point defects (GORELIK; LIFSHITS et al.) — vacancies and atoms in the space between lattice knots. During recovery electric resistance remains almost constant. It is connected with the rapid decrease of vacancies and simultaneous increase of the concentration of C and N atoms in α-Fe lattice.

In the temperature interval of collective recrystallization (500—700 °C) a grain growth occurs. It is connected with migration of high angle boundaries and partial annihilation of dislocations (GORELIK). At the same time concentration of C atoms in α-Fe lattice continues to increase. As it follows from Fe–N diagram (HOUDREMONT) the solubility of N atoms in α-Fe lattice increases most intensively to the temperature ~ 500 °C. Beginning from temperatures 580 °C the solubility of N in α-Fe decreases rapidly with further increase of temperature. Recrystallization developed at these temperatures is the reason for the Vickers hardness decrease. May be dislocations have enough energy to break from the atmospheres of N and C atoms that might arise on them. As it was mentioned above resistance is slightly sensitive to dislocations density. So the resistance values in the interval 500—700 °C increase considerably mainly as a result of increased solubility of C atoms.

Table 4
Relation between texture components {111} and {100} with respect to the annealing regime

Texture component	P_{hkl}, %				$P_{\{111\}}/P_{\{100\}}$			
	cold rolling	regime A	regime B	regime C	cold rolling	regime A	regime B	regime C
{111}	45	55	60	50	1.1	1.8	2.1	1.4
{100}	44	31	28	36				

After an observation of the decrease of lattice unit's volume at 500 °C (connected probably with decrease of residual stresses during recrystallization) its volume increases upon further annealing (probably as a result of increased solubility of C atoms).

In Table 4 the changes of texture components {111} and {100} during annealing with the three regimes are shown. It is obvious that the intermediate keepings at 500 °C and 550 °C increase the relation $P_{\{111\}}/P_{\{100\}}$ ($\bar{R}$ factor, respectively). The effect is more pronounced when the temperature of intermediate keeping is 550 °C.

4. Summary

The complex investigation of changes of some physical parameters through recrystallization annealing of a primary aged (by two months keeping at room temperature) cold rolled steel 08 killed with Al, permits us to draw conclusions about the influence of carbon and nitrogen atoms solved in the α-Fe lattice. Carbides and nitrides, getting dissolved during annealing, increase the concentration of C and N atoms in the α-Fe lattice. The summary concentration of these atoms will be the highest at the interval ~ 500 to 580 °C. Their retarding effect on recrystallization processes is relatively the most pronounced at this temperature interval. According to HUTCHINSON and IORDANOVA et al. nucleation in bcc metals begins earliest and proceeds more intensively in texture component {111} than in {100}. The isothermal keeping at 500—550 °C increases the concentration of C and N atoms predominantly in texture component {100} and smaller in more perfect at these temperatures component {111}. Impurities predominantly precipitate on dislocations, decrease the elastic energy existing around them and make slower their movement. As a result the tendency of more intensive developement of {111} texture component during recrystallization of this kind of steels, gets deeper. The effect is better expressed when the intermediate keeping is at 550 °C and the joint solubility of C, N, and Al atoms is higher than at 500 °C.

The results received show that influence of the atoms getting solved during recrystallization annealing of 08 killed with Al steel is qualitatively similar to that one, caused by disperse AlN precipitations.

In 08 with Al killed steel, for which C and N atoms are included in carbides and nitrides before recrystallization, the temperature of beginning recrystallization is moved by about 50 °C to the lower temperatures in comparison to those produced in ordinary conditions (with low rolling temperature of the hot rolled strip or without keeping at room temperature before annealing). Heating with a low speed, or with an intermediate keeping at 500 °C, improves recrystallization texture of 08 killed with Al steel in which carbides and nitrides have been completely separated before annealing.

References

DIAKONOVA, V. S., IVANOVA, G. L., UTEVSKI, L. M.: Physica metallov i metallovedenie (in russ.) **34**, 3 (1972)

DILLAMORE, I. L., SMITH, C. F. E., WATSON, T. W.: Met. Sci. J. **1**, vol. 3 (1966)

GORELIK, S. S.: Recrystallization in metals and alloys, Moscow 1980 (in russ.)

HOUDREMONT, E.: Handbuch der Sonderstahlkunde, Berlin 1956

HUTCHINSON, W. B.: Metal science, vol. 8, (1974)

IORDANOVA, I. et al.: Bulgarian J. Physics **9**, 1 (1982)

IORDANOVA, I., GORELIK, S. S., NOVIKOV, In.: Izvestia VUZ, Tchernaya Metallurgia **5** (1980), (in russ.)

LANCFORD, W. G., SNYDER, S. H., BAUSCHER, Y. A.: Trans. A.S.M., 42 (1950)

LIFSHITZ, B. G., KRAPOSHIN, W. S., LINETSKY, I. L.: Physical Properties of Metals, Moscow 1980 (in russ.)

MEYZAND, J. et al.: Texture and Properties of Materials, London 1976

MEYZAND, J., PARNIERE, P., THOMAS, B. f.: Mem. Sci. Rev. Met. 75, (1978)

PARNIERE, P.: Mater. of ICOTOM, 6 (I) — 1, Tokyo 1981

SALTIKOV, S. A.: Stereometricheskaya Metallographia, Moscow 1970 (in russ.)

(Received 1) April 4, 1987;
2) May 21, 1987)

Authors' addresses:

Dr. Ivanka Alexandrova IORDANOVA, Dr. Liubomir Borisov ILIEV
University of Sofia, Dept. of Solid State Physics
Faculty of Physics
5, Anton Ivanov Av.
1126 Sofia, Bulgaria

Dr. Vasil Spasov BOJINOV
University of Sofia, Faculty of Chemistry
Department of Inorganic Chemical Technology
1, Anton Ivanov Av.
1126 Sofia, Bulgaria

Book Review

W. T. LINDLEY (ed.)

Gallium Arsenide and Related Compounds 1986

Institute of Physics Conference Series 83
Proceedings of the Thirteenth International Symposium on Gallium Arsenide and Related Compounds held in Las Vegas, Nevada, 28 September – 1 October 1986

Institute of Physics, Bristol 1987, 594 Seiten, zahlreiche Abbildungen und Tabellen, Autorenindex, £ 45.00, ISBN 0-85498-176-4

Es ist beeindruckend, wie der Erkenntniszuwachs pro Jahr auf dem Gebiet der $A^{III}B^{V}$-Halbleiter voranschreitet. Die Arbeiten werden zielgerichtet auf die Entwicklung neuer Bauelementegenerationen focussiert. Dabei werden neue technologische und theoretische Konzepte in einer Tiefe und mit einem Aufwand ausgelotet, wie sie bisher nur vom Silizium bekannt sind. Dem tragen die drei Plenarvorträge (H. WATANABE, S. USUI: Atomic layers epitaxy; R. D. BURNHAM et al.: Impurity-induced disordering and its applications in optoelectronics; F. CAPASSO: Resonant tunneling transistors and superlattice devices by band-gap engineering) Rechnung. Auf dem Gebiet der Substratentwicklung konzentrieren sich die Bemühungen eindeutig auf halbisolierendes GaAs, wobei Reinheitsprobleme und Kristallperfektion im Vordergrund stehen. Bei Czochralski-InP ist es durch As-Dotierung gelungen, die Versetzungsdichte deutlich zu senken.

Bei den Epitaxie-Verfahren stehen nach wie vor die OMVPE und MBE-Verfahrensvarianten im Mittelpunkt. Neben der tiefgreifenden Untersuchung technologierelevanter Probleme ergeben sich auch interessante Aspekte von grundlegender Natur wie z.B. das Auftreten von chalcopyritartigen Phasen in Mischkristallen oder der Nachweis bisher unbekannter Defektzentren.

Die Beiträge über optoelektronische und Hochgeschwindigkeitsbauelemente nehmen den breitesten Raum ein. Das Spektrum erstreckt sich von neuen Aufbauprinzipien für GaInAsP/InP-Heterostrukturlaser (M. OSHIMA u. a.: Surface emitting lasers with slant edge mirrors and multilayer reflectors) bis zu einer GaAs/AlGaAs-Superlattice Photodiode (F. Y. JUANG u. a.)· Beeindruckend sind die Entwicklungen auf dem Sektor der GaAs-Elektronik, wobei der Bogen zum Plenarvortrag von F. CAPASSO geschlossen wird. Interessant ist auch ein Vorschlag für einen bipolaren InGaAs(P)/InP- Transistor, der von PANISH und TEMKIN gemacht wird und der sowohl für die Integration in faseroptischen Systemen als auch für die Hochgeschwindigkeitselektronik eine interessante Lösung darstellt.

Der Band schließt in Aufbau, Form und Qualität an seine Vorgänger an und spiegelt die Entwicklungstendenzen in aller Breite wider. Er ist für Materialentwickler genau so ergiebig, wie für Physiker und Bauelementefachleute.

G. KÜHN

Cryst. Res. Technol.	22	1987	12	1493–1496

C. Ascheron, H. Neumann

Sektion Physik, Karl-Marx-Universität Leipzig

Depth Profile of the Microhardness in Helium Implanted GaP

Depth profiles of the damage density and the microhardness are measured in GaP single crystals implanted with 1 MeV helium ions. From an analysis of the experimental data it follows that the microhardness increases up to a damage density of about 16% due to point defect hardening. At higher damage densities the microhardness decreases rapidly, probably due to the formation of extended defects.

Die Tiefenprofile der Defektdichte und der Mikrohärte wurden für mit 1 MeV Heliumionen implantierte GaP-Einkristalle gemessen. Aus einer Analyse der experimentellen Daten folgt, daß die Mikrohärte bis zu einer relativen Defektdichte von etwa 16% infolge von Punktdefekthärtung ansteigt. Bei höheren Defektdichten fällt die Mikrohärte steil ab, wahrscheinlich infolge der Bildung ausgedehnter Defekte.

1. Introduction

Ion implantation is finding increasing applications as a process to modify the mechanical properties of metals, ceramics and crystalline insulaters (Burnett, Page 1984b, 1985, 1986; Hioki et al.; Williams). Recent studies of the elastic properties and the hardness of ion implanted silicon (Burnett, Briggs; Burnett, Page 1984a, b) and of the microhardness of proton and helium implanted GaP (Ascheron, Neumann; Ascheron et al. 1986a) have shown that ion implantation has also profound effects upon the mechanical properties of semiconducting materials. Since these changes in the elastic and mechanical properties may influence the performance and the lifetime of semiconductor devices (Watts, Willoughby) a more detailed study of the implantation induced effects is required. In our previous studies the microhardness changes in the near-surface region of proton and helium implanted GaP single crystals have been investigated (Ascheron, Neumann; Ascheron et al. 1986a). It was the aim of the present work to determine the depth profile of the microhardness in helium implanted GaP single crystals the knowledge of which can be important in thin film device technologies.

2. Experimental

The samples used in the experiments were (111)-oriented GaP single crystals grown by the liquid encapsulation Czochralski technique. The crystals were doped with tellurium and had electron concentrations in the range from $2 \cdot 10^{17}$ to $5 \cdot 10^{17}$ cm^{-3} at room temperature. Irradiation was carried out with 1 MeV He^+ ions in a vacuum of about 10^{-4} Pa and with the samples held at room temperature. A constant helium ion fluence of $5 \cdot 10^{16}$ cm^{-2} was used in all cases.

To determine the depth profile of the damage density and the microhardness bevelled samples were prepared by sputtering with 1 keV Ar^+ ions as described by Ascheron et al.

(1985). The depth dependence of the radiation induced damage density was measured by the Rutherford backscattering channeling technique using 1.7 MeV He^+ ions (ASCHERON et al. 1986b). The hardness was determined with a Vickers microhardness tester. The depth profile of the microhardness was evaluated by corresponding changes in the position of the indenter on the surface of the bevelled samples. A load of 20 g was applied for a period of 15 s which yields indentation depths in the range from 0.5—0.7 μm. Immediately before the hardness measurements the samples were shortly dipped in aqua regia to remove the sputtering-induced surface damage and surface contaminations. To reduce the error in the hardness values twenty microhardness determinations were made at random for equivalent positions on the surface of the bevelled sample and the average taken. The standard deviation in the hardness values obtained in this manner was found to be about 5%.

3. Results and discussion

A typical example for the depth profile of the relative damage density N_d/N_0 and the microhardness ratio H/H_0 is shown in Figure 1. The microhardness H_0 of the unimplanted crystals was always determined separately and was found to be $H_0 = (7.6 \pm 0.1)$ GPa which is in good agreement with data published previously (ASCHERON et al. 1986a; KÜHN et al.). Qualitatively, the microhardness profile represented in Figure 1b agrees with measurements reported for ion implanted metals (PETHICA et al.; HAMAGUCHI et al.; SHIRAISHI, FUKAI). However, comparing the depth profiles of the damage density and of the microhardness it is obvious that their maximum values are located at different depths d from the sample surface (Fig. 1). From our experimental data it follows that the microhardness increases only up to a relative damage density of about $N_d/N_0 = 0.16$ whereas a rapid decrease of H is observed at higher damage densities. Now, in studying the implantation induced changes in the microhardness of the near-surface region in proton and helium im-

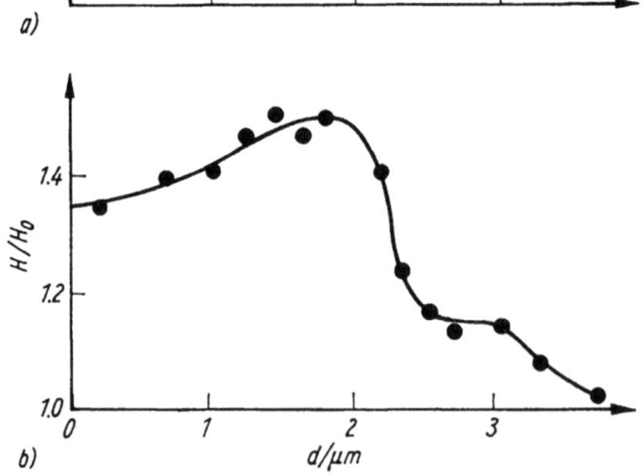

Fig. 1. Depth dependence of the relative damage density N_d/N_0 (a) and the microhardness ratio H/H_0 (b) for a GaP single crystal implanted with 1 MeV He^+ ions at a fluence of $5 \cdot 10^{16}$ cm^{-2}

planted GaP single crystals (ASCHERON, NEUMANN) it has been established that in the damage density range $N_d/N_0 < 0.1$ the microhardness H of the implanted surface follows the relation

$$H = H_0 + B(N_d/N_0)^{1/2}, \tag{1}$$

where the constant B is independent on the implanted species and equal to $B = 10.5$ GPa. Furthermore, it has been argued that the square root dependence of H on N_d can be ascribed to hardening due to point defects. If we plot the hardness ratios H/H_0 against $(N_d/N_0)^{1/2}$ taken from Figure 1 we see that relation (1) with the same constant B describes also the depth dependence of the microhardness up to $N_d/N_0 \approx 0.16$ (see Fig. 2). The maximum hardness obtained at this damage density is about 11.5 GPa. Thus, it is certainly justified to suppose that the microhardness change in helium implanted GaP is governed by point defect hardening for damage densities in the range $N_d/N_0 < 0.16$.

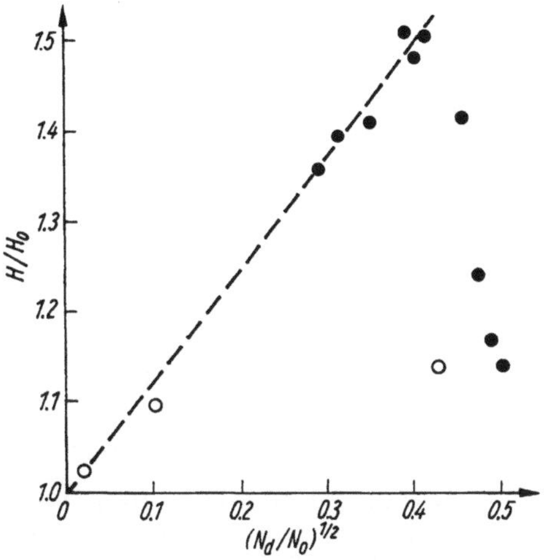

Fig. 2. Plot of the microhardness ratio H/H_0 against the square root of the relative damage density N_d/N_0. The dashed curve is calculated with relation (1), the open points correspond to depths d beyond the damage density peak

At higher damage densities the microhardness decreases rapidly down to values only slightly above the microhardness of the unimplanted crystals (Fig. 2). Probably, this effect is due to the formation of larger agglomerates of point defects or of extended defects like voids and dislocations as they have been found in helium implanted GaP single crystals for damage densities $N_d/N_0 > 0.15$ (ASCHERON et al. 1986b). It is interesting to note that recently reported results for the fluence dependence of the microhardness in ion implanted silicon, sapphire and some other materials (BURNETT, BRIGGS; BURNETT, PAGE 1984a, b, 1985, 1986; HIOKI et al.) confirm our measurements. In these investigations it has been found that the microhardness increases with the ion fluence at low fluences, then goes through a maximum at some critical fluence and decreases at higher fluences. If radiation induced amorphization occurs the microhardness becomes even lower than in the crystalline material. Thus the decrease of the microhardness of ion implanted materials above some critical fluence or damage density seems to be a general phenomenon. Similar effects have also been observed in ion implanted metals (WILLIAMS and references cited therein) In conclusion we note that in our measurements the microhardness emained always larger than in the unimplanted substrates (Fig. 1). This result corresponds to the experimental fact that at fluences below about 10^{17} cm^{-2} of 1 MeV helium ions no amorphization occurs in GaP single crystals (Ascheron et al. 1986b).

Summarizing our results it seems to be evident that at relative damage densities in the range $N_d/N_0 < 0.16$ the implantation induced change in the microhardness of helium implanted GaP single crystals follows relation (1) and can therefore be ascribed to point defect hardening. However, to elucidate the nature of the defects responsible for the microhardness increase further investigations are required including also other implanted species. The decrease in the microhardness at damage densities $N_d/N_0 > 0.16$ can certainly be ascribed to the influence of extended defects. However, both the nature of these defects and the mechanisms involved in the hardness change remain unsettled until now.

The authors are grateful to Prof. Dr. G. Dlubek and to Dr. R. Krause for their kind support in the experimental measurements.

References

Ascheron, C., Bauer, C., Sobotta, H., Riede, V.: phys. stat. sol. (a) **89**, 549 (1985)
Ascheron, C., Neumann, H.: Nucl. Instrum. Meth. B **18**, 161 (1987)
Ascheron, C., Neumann, H., Dlubek, G., Krause, R.: J. Mater. Sci. Lett. **5**, 891 (1986a)
Ascheron, C., Schindler, A., Flagmeyer, R., Otto, G.: phys. stat. sol. (a) **96**, 555 (1986b)
Burnett, P. J., Briggs, G. A. D.: J. Mater. Sci. **21**, 1828 (1986)
Burnett, P. J., Page, T. F.: J. Mater. Sci. **19**, 845 (1984a)
Burnett, P. J., Page, T. F.: J. Mater. Sci. **19**, 3524 (1984b)
Burnett, P. J., Page, T. F.: J. Mater. Sci. **20**, 4624 (1985)
Burnett, P. J., Page, T. F.: Inst. Phys. Conf. Ser. **75**, 789 (1986)
Hamaguchi, Y., Kuwano, H., Kamide, H., Miura, R., Yamada, R.: J. Nucl. Mater. **133/134**, 636 (1985)
Hioki, T., Itoh, A., Ohkubo, M., Noda, S., Doi, H., Kawamoto, J., Kamigaito, O.: J. Mater. Sci. **21**, 1321 (1986)
Kühn, G., Leonhardt, A., Hübner, K.: Kristall und Technik **7**, 1077 (1972)
Pethica, J. B., Hutchings, R., Oliver, W. C.: Nucl. Instrum. Meth. **209/210**, 995 (1983)
Shiraishi, K., Fukai, K.: J. Nucl. Mater. **117**, 134 (1983)
Watts, D. Y., Willoughby, A. F. W.: J. Appl. Physics **56**, 1869 (1984)
Williams, J. S.: Rep. Progr. Phys. **49**, 491 (1986)

(Received May 12, 1987)

Authors' address:

Dr. C. Ascheron, Dr. H. Neumann
Sektion Physik, Karl-Marx-Univesität
Linnéstraße 5
Leipzig
DDR-7010

Cryst. Res. Technol.	22	1987	12	1497–1503

C. ASCHERON, R. FLAGMEYER, G. OTTO, J. VOGT, H. U. FREY

Sektion Physik der Karl-Marx-Universität Leipzig,
WB Angewandte Kernphysik

Study of Radiation Damage on Proton-bombarded GaP by PIXE

On GaP single crystals bombarded with 0.3 MeV and 1.0 MeV protons the fluence and depth dependences of radiation damage are studied by means of particle induced X-ray emission spectrometry under chaneling conditions (PIXE-C) and the Rutherford backscattering/channeling technique (RBS-C). It is demonstrated that PIXE-C is suitable to study depth profiles of damage density by the combination with the bevelling technique.

An mit 0,3 MeV- und 1,0 MeV-Protonen bestrahlten GaP-Einkristallen werden die Dosis- und Tiefenabhängigkeiten der Strahlenschädigung mittels der teilcheninduzierten Röntgenemission unter Kanalisierungsbedingungen und der Rutherford-Rückstreu/ Kanalisierungstechnik untersucht. Es wird gezeigt, daß die Röntgenanregung unter Kanalisierungsbedingungen in Kombination mit der Schrägschlifftechnik Aussagen zu Tiefenprofilen von Strahlenschäden ermöglicht.

1. Introduction

PIXE has become an established method for trace element analysis (see e.g. Papers 3rd PIXE Conf., FOLKMANN; JOHANSSON, JOHANSSON). The applicability of PIXE to depth profiling measurements depends on the excitation and absorption of the X-rays in the matrix. Several attempts have been made to improve the depth resolution of PIXE (e.g. FELDMAN et al.; PABST; BENKA et al.; AHLBERG). Also, coincidence measurements of PIXE and RBS-C have been used (BAHIR et al.) in order to combine the high element resolution of PIXE with the good depth resolution of RBS. On crystalline material PIXE is also suitable to give information on the perfection of the target crystal if the protons are incident parallel to a lattice direction (LUERA). The aligned incidence reduces small-impact parameter processes (i.e., inner-shell ionization or direct backscattering) by a factor of about 10 (DAVIES et al.). Since both processes are similarly reduced, they respond alike to radiation damage (GERETSCHLÄGER; BONTEMPS et al.) or nonsubstitutional foreign atoms (BHATTACHARYA et al.; TAKAI et al.; CHEMIN et al.; VASILJEV et al.).

For lattice site location studies channeled PIXE (PIXE-C) (BHATTACHARYA et al.; TAKAI et al.; CHEMIN et al.) can be used like channeled particle induced γ-ray emission spectrometry (PIGE-C) (e.g. PICRAUX, VOOK). PIXE-C can be superior to PIGE-C and to RBS-C in special cases due to the restricted availability of nuclear resonance reactions in PIGE-C and due to the problems with impurities of similar or lower mass in RBS-C measurements.

A further advantage of the combination of PIXE with the channeling technique lies in getting more detailed information on the damage distribution over the sublattices of compound semiconductors with atoms of similar mass, e.g., GaAs (BHATTACHARYA) and InSb, where RBS-C is very sophisticated.

In this work PIXE-C and RBS-C measurements are used on proton bombarded GaP single crystals to observe the fluence and depth dependent damaging immediately during the process of H^+ implantation as well as to study the remaining damage.

2. Experiments

The samples investigated were Czochralski-grown (111) GaP single crystals chemico-mechanically polished. H^+ implantation was carried out with 0.3 MeV and 1.0 MeV protons.

RBS-C and PIXE-C measurements were made with 1 MeV proton and He^+ ion beams (projected ranges $R_p \approx 12.6$ µm and 2.6 µm, resp.). For the RBS-C and PIXE-C measurements the protons and He^+ ions were incident parallel to the ⟨111⟩ direction. During implantation the crystal was tilted 10° off the ⟨111⟩ axis. Irradiation was carried out at room temperature with fluences ranging from 10^{16} cm^{-2} to 1.7×10^{18} cm^{-2} on the van de Graaff accelerator of the Karl Marx University, Leipzig. On the sample implanted with 0.3 MeV protons the surface was bevelled by 1 keV Ar^+ ion sputtering ($\tan \alpha = 2 \times 10^{-4}$).

The X-ray yields were detected with a Si (Li) detector (FWHM ≈ 190 eV) shielded against the vacuum by an 6 µm Mylar foil and located at 135° with respect to the beam direction.

3. Results and discussion

The damage was observed by simultaneously measuring the yields of backscattered protons or He^+ ions and the yields of P–K (2.02 keV) and Ga–Kα (9.25 keV) X-rays.

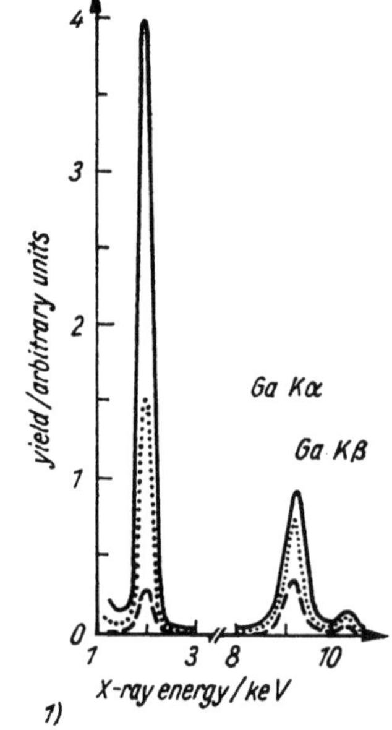

Fig. 1. Energy spectra of the P – K and Ga–K X-rays excited by 1 MeV proton irradiation of (111) GaP in a random and in the ⟨111⟩ lattice direction. ——— random, ... saturation, – – – virgin

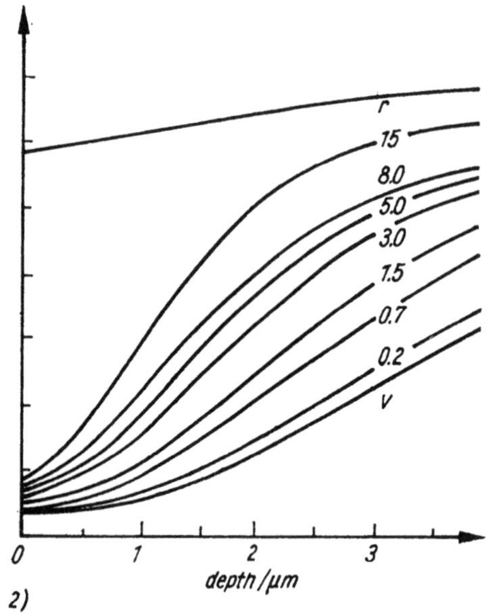

Fig. 2. Depth dependence of the backscattering yield of (111) GaP irradiated with several fluences of 1 MeV protons under room temperature conditions (the P spectrum which is shifted in energy against the Ga spectrum was converted into depth, v – virgin sample; r – random incidence (corresponding to total disorder)). $D/10^{17}$ · cm^{-2}

Figure 1 shows the X-ray spectrum of GaP obtained by excitation with 1 MeV protons along a random and the ⟨111⟩ direction. The corresponding depth dependent backscattering spectra and one typical depth profile of damage density are given in Figures 2 and 3. Since in the energy dependence of the backscattering spectrum of compound material two depth scales shifted relative to each other are necessary, in Figure 2 the two depth scales were brought into accordance by taking into account the shift of the P- against the Ga spectrum.

The relation in intensity of the P–K and Ga–Kα radiations is determined by a) the energy dependent origination of X-rays and the self-absorption in the material, b) the attenuation in the Mylar foil, in approximately 10 mm air, and in the Be window of the detector as well as by the detector efficiency.

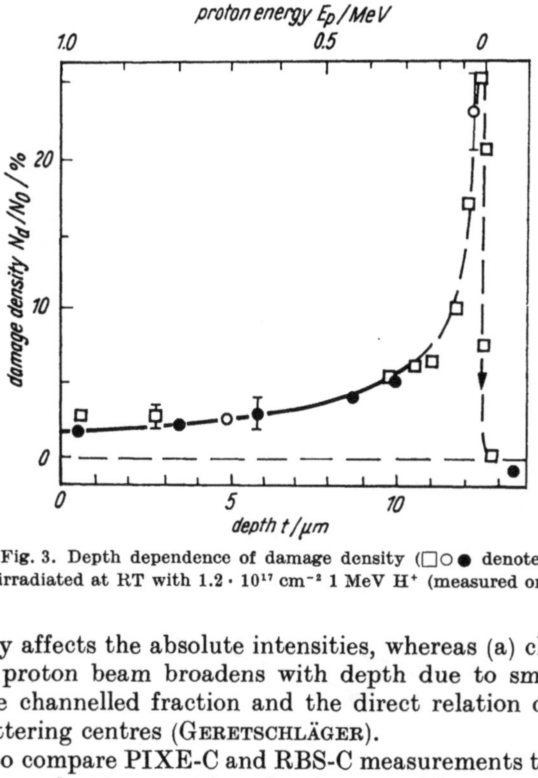

Fig. 3. Depth dependence of damage density (□○● denote different samples) of GaP (111) irradiated at RT with $1.2 \cdot 10^{17}$ cm^{-2} 1 MeV H$^+$ (measured on a bevelled sample (ASCHERON)

(b) especially affects the absolute intensities, whereas (a) characterizes the analyzing depth. The proton beam broadens with depth due to small angle scattering, thus reducing the channelled fraction and the direct relation of X-ray yield to damage induced scattering centres (GERETSCHLÄGER).

In order to compare PIXE-C and RBS-C measurements the depth dependent changes of X-ray production must be taken into consideration.

The X-ray yield is proportional to the following expression:

$$\chi \sim \int_0^{R_p} \sigma_x(E(t))\, \chi_{\min}(t)\, e^{-\mu t}\, dt\,. \tag{1}$$

$\int_0^{R_p} \chi_{\min}(t)\, dt$ is the integrated dechannelling flux over the penetration range R_p, $e^{-\mu t}$ represents the self absorption and $\sigma_x(E(t))$ the cross section for X-ray production. For the channelled fraction of the proton beam the energy loss was assumed to be half that of the random case, roughly corresponding to that of ref. (SATTLER, VOOK). Figs. 4a and b show the depth dependent differential yields detectable on the surface $dY(t) = \sigma_x(E(t)) \cdot e^{-\mu t} \cdot \chi_{\min}(t)$ for the P–K and Ga–Kα X-rays, resp. (with $\sigma_x(E)$

(BRANDT, LAPIEKI) and μ (THINH, LEROUX)). $\chi_{\min}(t)$ was taken from Figure 2. For details of the calculation see ref. (FREY et al.).

As can be seen from the figures, about 50% of the P–K X-ray yields originate within the first 0.5 µm of the target whereas the corresponding depth is about 3 µm for the Ga–Kα X-ray yield. By comparing the X-ray curves it follows that the relative contribution of deeper layers to the measured X-ray yields is the higher the lower the dechanneling in the near-surface layer. Therefore, the mean analyzing depth is shifted to smaller depths with increasing damage density. This shift is more pronounced for the Ga–Kα radiation than for the stronger selfabsorbed P–K X-rays (nearly from 3.5 µm to 2.5 µm and from 1.0 µm to 0.5 µm for the Ga–Kα and P–K radiation, resp.).

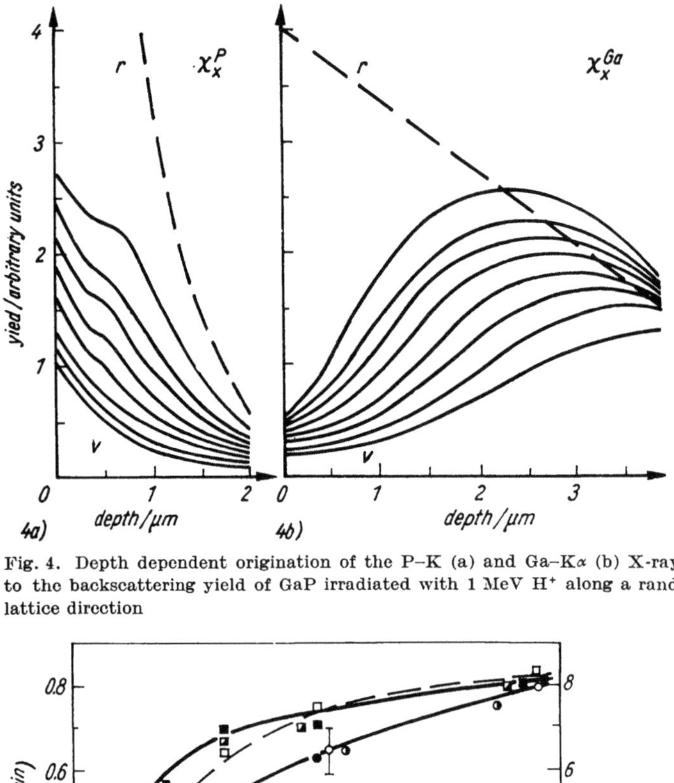

Fig. 4. Depth dependent origination of the P–K (a) and Ga–Kα (b) X-ray yields in relation to the backscattering yield of GaP irradiated with 1 MeV H^+ along a random and the ⟨111⟩ lattice direction

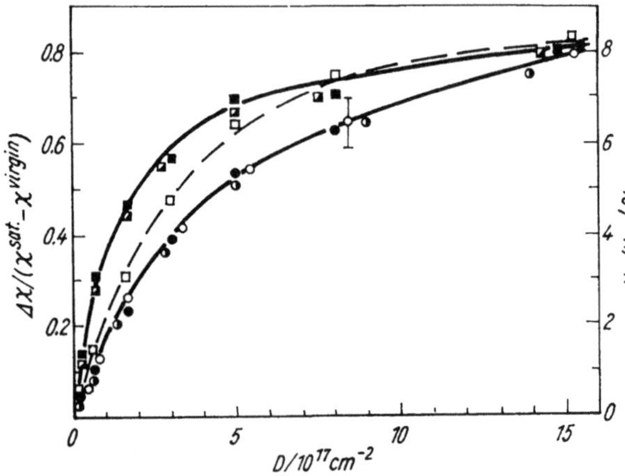

Fig. 5. Measured (filled symbols) and computed (half-filled symbols) P–K (●, ◑) and Ga–Kα (■, ◪) PIXE-C yields of (111) GaP as a function of the irradiated 1 MeV H^+ fluence in comparison to the mean damage densities within the layer to be analyzed □ N_d/N_0 (t = 1 µm); ○ N_d/N_0 (t = 1 µm)

Due to the strongly depth-dependent dechanneled fraction of the proton beam this shift cannot be neglected (especially if the Ga–Kα yield is monitored) in studies of radiation damage, even if the damage density does not strongly vary with depth (see Fig. 3).

The normalized fluence dependent changes in the Ga–Kα and P–K PIXE-C yields (values measured and computed by integrating the differential yields given in Figure 4)

$$\Delta\chi^{\rm rel}(D) = (\chi(D) - \chi^{\rm virg})/(\chi^{\rm sat} - \chi^{\rm virg}) \tag{2}$$

are shown in Figure 5. $\chi^{\rm virg}$ and $\chi^{\rm sat}$ are the yields of the undamaged crystal and at saturation of damage, resp. For comparison, Figure 5 contains the mean damage

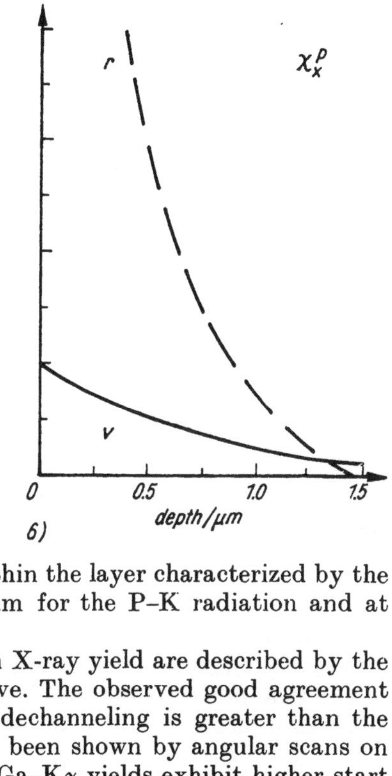

Fig. 6. Depth dependent origination of the P–K X-ray yield detectable at the surface for randomly (——) and channeled (– – –) incident 1 MeV He⁺ ions on (111) GaP

densities (determined by RBS-C measurements) within the layer characterized by the PIXE-C measurements, i.e., at about (0.5—1.0) µm for the P–K radiation and at about (2.5—3.5) µm for the Ga–Kα radiation.

As can be seen, the measured relative changes in X-ray yield are described by the relation to the backscattering yield discussed above. The observed good agreement implies that the critical distance of approach for dechanneling is greater than the maximum parameter for K shell ionization as has been shown by angular scans on GaP (Davies et al.) and GaAs (Price et al.). The Ga–Kα yields exhibit higher start and saturation values (45% and 80%, resp., of the random yield) than the P–K yields (7% and 32%, resp.). This and the stronger fluence dependent changes of the Ga–Kα X-ray yields, which are mainly attributed to ionization at greater depths by dechannelled protons, are due to the weaker absorption. Therefore, the changes in Ga–Kα yield cannot directly be related to damage density.

In addition to small impact parameter processes within the near-surface layer the Ga–Kα yields include a higher contribution resulting from dechanneling in this layer, leading to small impact parameter processes in deeper layers.

Consequently, the changes in Ga–Kα PIXE-C yield are more strongly indicative of low damage densities and strain around defects, which lead to dechanneling rather than the direct backscattering, than this are changes in the P–K yield. The fluence dependent changes in damage density and P–K yield agree qualitatively since the

P–K yields originate mainly within a depth region where the directly backscattered and the dechanneled fractions of the proton beam are proportional to each other (factor of about 2 for $N_d/N_0 \leqq 10\%$). This relation between damage density and PIXE-C yield is the closer the thinner the characterized layer.

We made some attempts to investigate depth profiles of damage density be means of PIXE-C. For this purpose the damage profile was revealed by bevelling the implanted sample. Depth dependent PIXE-C measurements were made by scanning the bevelled surface with the analyzing beam. In order to improve the depth resolution of the method 1 MeV He^+ ions were used for excitation. The corresponding mean

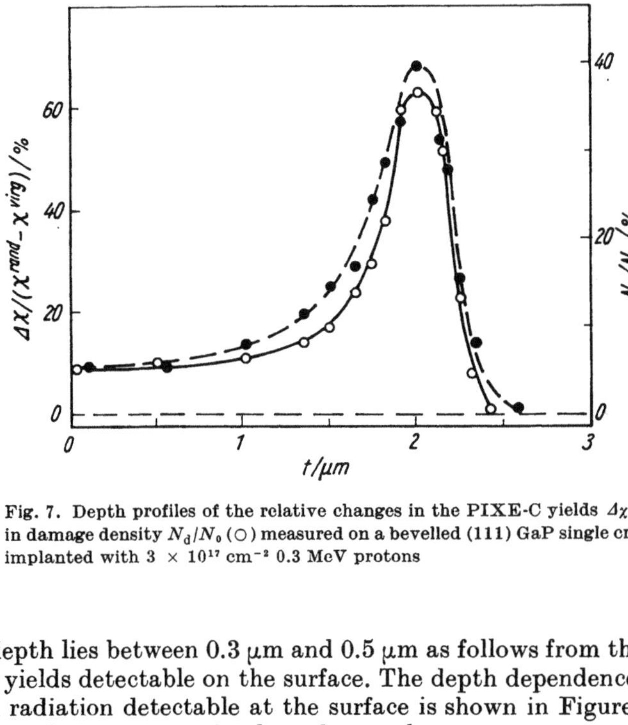

Fig. 7. Depth profiles of the relative changes in the PIXE-C yields $\Delta\chi/(\chi^{rand} - \chi^{virg})$ (●) and in damage density N_d/N_0 (○) measured on a bevelled (111) GaP single crystal (tan α = 2×10^{-4}) implanted with 3×10^{17} cm^{-2} 0.3 MeV protons

analyzing depth lies between 0.3 µm and 0.5 µm as follows from the depth dependent differential yields detectable on the surface. The depth dependence of the origination of the P–K radiation detectable at the surface is shown in Figure 6 for random and channelled incidence on an unimplanted crystal.

The shift of the mean analyzed depth towards the surface with decreasing fraction of channeled ions, which can be recognized in Figures 6 and 4, effects a better depth resolution within the region of the damage peak.

The depth dependent relative changes in the PIXE-C yield are presented in Figure 7 together with the depth profile of damage density.

Figure 7 shows that the profile of the relative changes in the PIXE-C yields largely matches the damage profile. Even the relatively sharp damage peak can be resolved. As was discussed above for excitation with protons, the relative changes in the X-ray and backscattering yields are directly related to each other. They exceed the changes in mean damage density by a factor of about 2. The measurements show that this relation holds over a wide range of damage densities (checked up to $N_d/N_0 = 40\%$). With this knowledge the relative change in the P–K PIXE-C yield can directly be transformed into damage density data. Thus PIXE-C measurements can directly be used for investigations of radiation damage.

4. Summary

Summarizing it can be stated that PIXE-C measurements give information on radiation damage. They reflect the backscattering yield of a certain depth interval in a modified form. The X-ray yield increases more strongly with fluence than the scattering centre density as also dechanneled and not only directly backscattered ions excite X-rays. The relation between the X-ray yield and damage density is the closer the thinner the layer characterized by the radiation.

The combination of PIXE-C with the ion beam etching technique offers the possibility of directly using PIXE-C for damage profiling.

The authors wish to thank Dr. A. Schindler for bevelling the samples.

References

AHLBERG, M.: Nucl. Instr. and Meth. **131**, 381 (1975)
ASCHERON, C., SCHINDLER, A., OTTO, G.: phys. stat. sol. **92**, 169 (1985)
BAHIR, G., KALISH, R., TSERRUYA, I.: Nucl. Instr. and Meth. **168**, 227 (1980)
BHATTACHARYA, R. S., RAI, A. K., YEO, Y. K., PRONKO, P. P., LING, C. S., WILSON, S. R., PARK, Y. S.: J. Appl. Physics **54**, **2329** (1983)
BHATTACHARYA, R. S.: J. Appl. Physics **54**, 5039 (1983)
BENKA, O., GERETSCHLÄGER, M., PAUL, H.: J. Appl. Physics **47**, 5090 (1976)
BONTEMPS, A., LIGEON, E., DANIELOU, R.: Rad. Eff. **22**, 195 (1974)
BRANDT, W., LAPIEKI, G.: Phys. Rev. A **23**, 1717 (1981)
CHEMIN, J. F., MITCHELL, I. V., SARIS, F. W.: J. Appl. Physics **45**, 537 (1974)
DAVIES, J. A., ERIKSSON, L., JOHANSSON, N. G. E., MITCHELL, I. V.: Phys. Rev. **181**, 548 (1969)
FELDMAN, L. C., POOLE, J. M., ERMANTS, F., SCHWARIZ, B.: Thin Solid Films **19**, 81 (1973)
FOLKMANN, F.: J. Phys. **E 8**, 429 (1975)
FREY, H., VOGT, J., OTTO, G.: J. Radioanal. Nucl. Chem. **99**, 193 (1986)
GERETSCHLÄGER, M.: Nucl. Instr. and Meth. **203**, 503 (1982)
JOHANSSON, S. A. E., JOHANSSON, T. B.: Nucl. Instr. and Meth. **137**, 473 (1976)
LUERA, T. F.: J. Appl. Physics **5**, 5792 (1980)
PABST, W.: Nucl. Instr. and Meth. **14**, 143 (1975)
Papers of the 3.rd PIXE Conf., Heidelberg (1983), in Nucl. Instr. and Meth. **B 3** (1984)
PICRAUX, S. T., VOOK, F. L.: Phys. Rev. Lett. **33**, 1216 (1974)
PRICE, P. B., GADEKEN, L. L., MAK, H. B., EWAN, G. T.: Rad. Eff. **39**, 205 (1978)
SATTLER, A. R., VOOK, F. L.: Phys. Rev. **161**, 244 (1967)
TAKAI, M., ISHIDA, T., GAMO, K., MAZUDA, K., NAMBA, S., MIZOBUCHI, A.: Jap. J. Appl. Physics **16**, 1647 (1977)
THINH, T. P., LEROUX, J.: X-Ray Spectrom. **8**, 85 (1979)
VASILJEV, V. K., GORSCHKOV, O. M., DANILOV, Yu. A., TULOVCHIKOV, V. S.: Fisika i Tekhnika Poluprov. **18**, 189 (1984)

(Received 1) April 13, 1987;
2) May 12, 1987)

Authors' address:

Dr. sc. Claus ASCHERON, Dr. R. FLAGMEYER, Prof. Dr. sc. G. OTTO,
Dr. J. VOGT, Dr. H. U. FREY
Sektion Physik der Karl-Marx-Universität
Linnéstr. 5
Leipzig
DDR-7010

Book Review

Der Maschinenschaden

Herausgegeben von der Allianz-Versicherungs-AG

Berlin und München, Preis: Einzelheft 17 DM, Jahresabonnement 102 DM, ISSN 0025-4517

Mit dem Heft 1/87 begann die Zeitschrift „Der Maschinenschaden" ihren 60. Jahrgang. Diese Publikation, die seit einiger Zeit mit dem Untertitel „Fachzeitschrift für Risikotechnologie" in einer Auflagenhöhe von 7000 Exemplaren erscheint, leistet einen wichtigen Beitrag zum internationalen Erfahrungsaustausch über die Aufklärung und Verhütung von Schadensfällen. Die Originalarbeiten überstreichen ein breites Spektrum, das von der schadenssicheren Dimensionierung über Aspekte der Fertigung, Prüfung und Betriebsüberwachung von Maschinen und Anlagen bis zu Fragen des Rechts reicht. Im angemessenen Umfang werden auch aktuelle Ergebnisse der werkstoffwissenschaftlichen Schadensforschung vorgestellt.

Diese thematische Vielfalt kommt auch im Heft 1/87 zum Ausdruck. Es wird u. a. über das betriebsfeste Dimensionieren von Schraubenverbindungen unter schwingender Beanspruchung, die Schwingfestigkeit von Kunststoffen, Erfahrungen mit mobilen Härteprüfgeräten und den Einsatz moderner Verfahren der Werkstoffprüfung zur Abschätzung der Restlebensdauer von Bauteilen unter dem Einfluß von Kriechen berichtet.

Berichte über Schadensfälle, Tagungsberichte sowie eine Bücher- und Schriftenschau erhöhen den Informationsgehalt dieser auch äußerlich sehr ansprechend gestalteten Fachzeitschrift.

H. Blumenauer

Cryst. Res. Technol.	22	1987	12	1505–1509

E. BOROŃSKI

Institute for Low Temperature and Structure Research
Polish Academy of Sciences, Wrocław, Poland

Studies of Positron Annihilation at Metal Vacancies with Two-component Positron-Electron Density Functional Theory[1])

Recently, the fully self-consistent two-component density-functional theory for electron-positron systems has been developed. The scheme is applied here in further calculations of positron annihilation in metal vacancies. In particular, angular correlation of positron-electron annihilation photons for Al vacancy is calculated and compared with conventional result. As important in some cases, the influence of self-interaction corrections on the results is discussed. Besides, the positron binding energies to vacancies containing H and He are calculated. They are considerably different then ones obtained in one-component approach.

Kürzlich wurde eine volle selbstkonsistente Theorie des Zweikomponenten-Dichte-Funktionals entwickelt. Im Rahmen dieser Theorie werden detaillierte Rechnungen über Positronen-Annihilation in Metallen mit Vakanzen präsentiert. Die Winkelkorrelationen der Photonen der Positron-Elektron-Annihilation im Al mit Vakanzen werden eingehend berechnet und mit den konventionellen Resultaten verglichen. Der in manchen Fällen wichtige Einfluß der Korrekturen der Selbstwechselwirkung auf die Ergebnisse wird diskutiert. Außerdem werden die Bindungsenergien der Positronen mit den Vakanzen, welche H und He enthalten, berechnet. Es besteht ein erheblicher Unterschied zwischen diesen und jenen, die mittels der Einkomponenten-Theorie berechnet worden sind.

An understanding of positron behaviour in defected solids requires a solution of a problem how the positron distorts the electronic structure of the system. In most calculations of positron characteristics in imperfect solids, or more generally, in systems where a positron can be localized, conventional approach bases on finding positron states treating electron density profiles as unperturbed by positron-electron attraction.

Recently, the full self-consistent two-component density-functional theory (TC-DFT) for electron-positron systems has been developed (BOROŃSKI, NIEMINEN) and successfully used in calculations of electron and positron densities in metal vacancies. This approach is further tested below for model system (vacancy in Al in the jellium approximation) by calculating the influence of self-interaction corrections on effective potentials and positron characteristics. Moreover, TC-DFT results for angular correlations of positron-electron annihilation photons are compared to conventional DFT data. Finally, TC-DFT is applied for studying positron binding to vacancies containing impurities such as H and He.

The scheme of TC-DFT calculations bases on the following formulas: total energy of system of interacting electrons and positrons can be written as the following functional (NIEMINEN)

[1]) Contributed paper presented during the European Meeting on Positron Studies of Defects, March 23–27, 1987, Wernigerode, G.D.R.

$$E[n_+, n_-] = E[n_+] + E[n_-] + \int dr\ V_{\mathrm{ex}+}(r)\,[n_-(r_-) - n_+(r)] - \\ - \int dr \int dr' \frac{n_-(r)\,n_+(r)}{|r - r'|} + E_c^{e-p}[n_+, n_-]\,. \tag{1}$$

According to Kohn-Sham prescription one can write a set of equations for electrons and positrons

$$-\frac{1}{2}\nabla^2\Psi_i(r) + \left(\mu_{\mathrm{XC}}(n_-(r)) + \Phi(r) + \frac{\delta E_c^{e-p}[n_+, n_-]}{\delta n_-(r)}\right)\Psi_i(r) = \varepsilon_i\Psi_i(r)\,, \tag{2a}$$

$$-\frac{1}{2}\nabla^2\Psi_i^+(r) + \left(\mu_{\mathrm{xc}}(n_+(r)) - \Phi(r) + \frac{\delta E_c^{e-p}[n_+, n_-]}{\delta n_+(r)}\right)\Psi_i^+(r) = \varepsilon_i^+\Psi_i^+(r)\,, \tag{2b}$$

where μ_{xc} is the exchange-correlation potential

$$\Phi(r) = \int dr' \frac{n_-(r') - n_+(r') - n_0(r')}{|r - r'|}\,, \tag{3}$$

is the Hartree-Coulomb potential, and

$$n_-(r) = \sum_{\varepsilon_i < \varepsilon_F} |\Psi_i(r)|^2\,, \qquad n_+(r) = \sum_i^{N_+} |\Psi_i^+(r)|^2 \tag{4}$$

are electron and positron densities, respectively. The n_0 is the density of positive charge corresponding to an external potential. Electron and positron potentials are the following

$$\mu_i(n_+, n_-) = \frac{\partial E_v^{e-p}(n_+, n_-)}{\partial n_i} \qquad (i = -, +)\,. \tag{5}$$

where

$$E_v^{e-p}(n_+, n_-) = n_<[a(n_>) + b(n_>)\,n_< + c(n_>)\,n_>^2]$$

with

$$\begin{aligned} a(n) &= \varepsilon_{\mathrm{AP}}(n)\,, \\ b(n) &= [3\varepsilon_{\mathrm{L}}(n) - 2\varepsilon_{\mathrm{AP}}(n) - \mu_0(n)]/n\,, \\ c(n) &= [\mu_0(n) + \varepsilon_{\mathrm{AP}}(n) - 2\varepsilon_{\mathrm{L}}(n)\ /n^2\,. \end{aligned} \tag{6}$$

$\varepsilon_{\mathrm{AP}}$, ε_{L}, μ_0 are correlation energies per particle and appropriate derivaties for one positron in electron gas and for compensated electron-positron plasma (see Boroński, Nieminen).

A problem has existed since papers on TC-DFT appeared (Chakraborty, Siegel; Nieminen): Can we use TC-DFT approach for the case of one positron in electron gas and what is the difference between the TC-DFT corrected with respect to positron self-interaction effects (SIC) and uncorrected one? From the formal point of view, there is an inconsistency in Chakraborty's paper; even not taking into account the lack of full TC positron-electron correlation positron is treated in some cases as a particle in other ones considered as density distribution of positive charge.

In papers (Nieminen, Boroński) and (Nieminen) the problem is treated consequently in TC-DFT from the very beginning and even in one-positron limit to the positron is considered as a cloud of positive charge with ferromagnetic type of exchange and correlation. To answer the question about the meaning of self-interaction effects electron and positron states in Al vacancy are calculated in TC-DFT approach with included and excluded self-interaction. Following initial calculations of BOROŃSKI and NIEMINEN the SI-correction in Schrodinger-like equation for positron was taken as

$$V_{\text{eff}}(r) = -\int \frac{n_+(r')}{|r - r'|} dr' - \mu_{\text{xc}}(n_+(r)) . \tag{7}$$

As was shown by BOROŃSKI and NIEMINEN effective potential calculated in TC-DFT is much deeper and more long ranged than one obtained in conventional DFT approach. In turn, when comparing SIC-TC-DFT to uncorrected result of TC-DFT with respect to self-interaction effects, we can see it as only slightly deeper and longer (Fig. 1). It is interesting that such result is obtained from very different contributions to the effective potential in both cases. In non-SIC approach the tail of the potential comes mainly from self-exchange and self-correlation contributions but in SIC case the

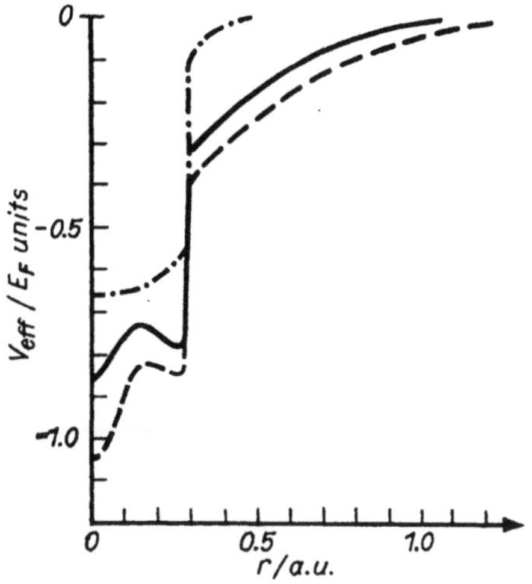

Fig. 1. Effective potentials experienced by the positron trapped in Al vacancy. Dashed-dotted curve represents conventional calculations. Solid curve corresponds to TC-DFT including self-interaction. Dashed curve is the result of TC-DFT corrected with regard to positron self-interaction

potential is rebuilt in the self-consistent procedure in such way that Hartree-Coulomb interaction itself can yield again very similar tail. As a result, binding energies and density profiles are also very similar, so there is almost no difference in positron localization in the defect for both cases. We can conclude that one-positron limit in TC-DFT despite some problems which it can create from formal point of view seems to be a reasonable approximation since the SIC and non-SIC results are almost the same qualitatively and quantitatively.

The next task was to compare the momentum densities of annihilation pairs calculated with TC-DFT and conventional density functional approach. Annihilation rates were calculated in paper of BOROŃSKI and NIEMINEN and results for both approaches were very close to each other. It is known, however, that these quantities are integrals over positron and electron densities and then the effects of diminished positron and increased electron densities inside a vacancy are cancelled. In partial annihilation rates nonlocal effects have usually greater meaning. Thus one is interested to what extent they manifest themselves in partial annihilation rate curves. The following approximation has been used for calculating momentum density distribution of annihilation quanta

$$R(p) \sim \sum_i |\int dr\, e^{-ipr} \Psi_i(r)\, \Psi_+(r) \sqrt{g(n_+(r), n_-(r))}|^2 , \tag{8}$$

where $g(n_+, n_-)$ is contact density depending on positron and electron densities (see BOROŃSKI, NIEMINEN). Results are presented in Figure 2. We can see that the differences are not very much. The curve corresponding to TC-DFT is somewhat higher for interesting values of momentum. One can expect, however, more visible differences if the square of contact density which is a global quantity is replaced by some kind of momentum dependent enhancement factor depending additionally on both electron and positron densities. Anyway, angular correlation spectra are much more sensitive to the details of electron-positron overlap then positron lifetime.

A general result obtained by BOROŃSKI and NIEMINEN was a greater positron binding energy calculated in TC-DFT in comparison to one-component DFT. To some extent it can be explained by a transfer of negative charge into the vacancy caused

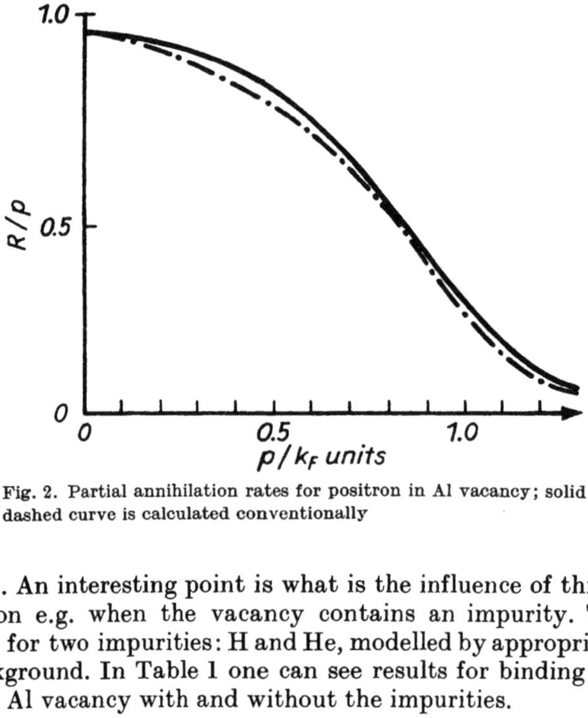

Fig. 2. Partial annihilation rates for positron in Al vacancy; solid curve is result of TC-DFT, dashed curve is calculated conventionally

by positron. An interesting point is what is the influence of this effect in more extremal situation e.g. when the vacancy contains an impurity. The calculations were carried out for two impurities: H and He, modelled by appropriately changed positive charge background. In Table 1 one can see results for binding energies and positron lifetimes in Al vacancy with and without the impurities.

Table 1
Positron binding energies and lifetimes in
a) conventional DFT, b) TC-DFT approach

	a	b	
vacancy	$E = 1.88$ eV	$E = 2.46$ eV	$t = 241$ psec
vacancy + H	$E = 1.56$ eV	$E = 2.93$ eV	$t = 195$ psec
vacancy + He	$E = 1.23$ eV	$E = 1.81$ eV	$t = 184$ psec

The first interesting feature is greater binding energy to H-vacancy complex than to the pure vacancy. This is in conflict with the result of (JENA, PONNAMBALAM) who found results of that type only for greater vacancy clusters. They, however carried out the conventional DFT calculations. It means that impurity ion may, owing to transferring additional negative charge into the vacancy, enhance positron-electron correlations so considerably that an additional attractive potential arises. This seems to be a general trend although other mechanisms or stronger Coulomb potential can diminish it as in e.g. He-vacancy complex.

References

BOROŃSKI, E., NIEMINEN, R. M.: Phys. Rev. **B34**, 3820 (1986)
CHAKRABORTY, B., SIEGEL, R. W.: Phys. Rev. **B27**, 4535 (1983)
NIEMINEN, R. M.: in: Positron Solid State Physics, ed.: W. BRANDT and A. DUPASQUIER, Amsterdam 1983
JENA, P., PONNAMBALAM, M. J.: Phys. Rev. **B26**, 5264 (1982)

(Received June 16, 1987)

Author's address:

Dr. E. BOROŃSKI
Institute for Low Temperature and Structure Research PAN
P.O. Box 937
50-950 Wrocław, Poland

Book Review

D. Stoyan, W. S. Kendall, J. Mecke

Stochastic Geometry and its Application

Mathematische Monographien, Band 69

Akademie-Verlag, Berlin 1987, with 83 figures and 18 tables; 345 pages. Price: Cloth M 50. – ISBN 3-05-500032-3

This volume is a revised and considerably extended version of the original German edition by Stoyan and Mecke reviewed in Cryst. Res. Technol. **11**, 1432 (1984). The author's aim is to acquaint not only mathematicians but also scientists and engineers with the results and methods of stochastic geometry. In the present version there is space for redundancy in representation which is of importance to the non-specialist. It was a good idea to reorganize the book in such a way as to begin with the elementary models such as the Poisson point process and the Boolean model, preparing the later developed general case, which culminate in the random measures. In this way, the reader will be introduced gently to the mathematical apparatus needed for stochastic geometry and its application. But most of the extension consists of a new chapter about random processes of geometrical objects written by W. S. Kendall and an interesting historically oriented foreword by D. G. Kendall, as well as of a lot of new results about stereology and statistical procedures. These results were obtained in the last three years by members of the school of stochastic geometry at the Mining College in Freiberg and at the University of Jena and are published in a monograph for the first time.

H. J. Girlich

Cryst. Res. Technol.	22	1987	12	1511–1514

J. GONDZIK, H. STACHOWIAK

Polish Academy of Sciences, Institute for Low Temperature and Structure Research, Wrocław

Positron Lifetime in Inhomogeneous Metallic Materials[1])

A positron is considered in an inhomogeneous jellium model according to a formalism following from the theory of liquids. This leads to a nonlocal treatment of the screening of this particle. Annihilation rates have been found inferior to the ones for a homogeneous electron gas, in agreement with positron lifetime measurements.

1. Introduction

Most theories describing electron-positron interaction in metals assume the jellium model (KAHANA; ARPONEN, PAJANNE; SJÖLANDER, STOTT; RUBASZEK, STACHOWIAK; KALLIO et al.; GONDZIK, STACHOWIAK). Real metals differ from jellium no matter whether they form pure perfect crystals or emphasis is put on vacancies, defects or impurities. Attempts to study $e^+ - e^-$ interaction in real metals have a long history (CARBOTTE, SALVADORI; BONDERUP et al.; CHAKRABORTY; FUJIWARA; SORMANN, PUFF). Let us mention the approach used by DANIUK et al., where the annihilation characteristics are treated in the local density approximation.

This work is a development of the liquid theory approach proposed previously by the authors (GONDZIK, STACHOWIAK). It allows to treat $e^+ - e^-$ interaction in an inhomogeneous medium without using the local density approximation. This makes possible as well to check the results of the local density approximation as to go beyond it in actual computations.

2. Basic formulas

The approach bases on the theory of liquids according to which for jellium the trial wave-function of the electron-positron system is written in the Jastrow-type (KALLIO et al.; PIETILÄINEN, KALLIO):

$$\Psi = \prod_{i=1}^{N} R(r_i - r) \prod_{i<j}^{N} f(r_i - r_j)\, \Phi\,, \tag{1}$$

where R and f play the role of variational parameters, r_i and r mean the coordinates of an electron and the positron respectively and Φ is a Slater determinant of plane waves which makes Ψ antisymmetric. After applying the variational principle the following Euler-Lagrange equation has been derived for the screening cloud density:

$$-\frac{1}{2\mu} \nabla^2 \sqrt{\varrho(r)} + [V(r) + W(r)] \sqrt{\varrho(r)} = 0\,, \tag{2}$$

$$\mu = \frac{m \cdot M}{m + M}, \tag{2a}$$

[1]) Contributed paper presented during the European Meeting on Positron Studies of Defects, March 23–27, 1987, Wernigerode, G.D.R.

where $V(r)$ is the electron-impurity potential, m and M are the electron and impurity masses, respectively. $W(r)$ is taken from the Hartree-Fock approach in the form (GONDZIK, STACHOWIAK):

$$W(r) = \int \frac{\varrho(r') - \varrho_0}{|r - r'|} d^3r' + V_{xc}[\varrho(r)], \tag{3}$$

instead of the form obtained by KALLIO et al.. The exchange-correlation potential V_{xc} is taken in the form proposed by HEDIN and LUNDQVIST in the local approximation.

Eq. (2) has now the following form:

$$-\frac{1}{2\mu} \nabla^2 \sqrt{\varrho(r)} + \left[-\frac{1}{r} + \int \frac{\varrho(r') - \varrho_0}{|r - r'|} d^3r' + V_{xc}\right] \sqrt{\varrho(r)} = 0\,. \tag{4}$$

In the above equation $\varrho_0 = 3/4\pi r_s^3$ is the electron density of jellium. In fact, however, in real metals there is no positive charge in the interstitial space where the positron is expected to be found. In order to take into account the crystal structure more

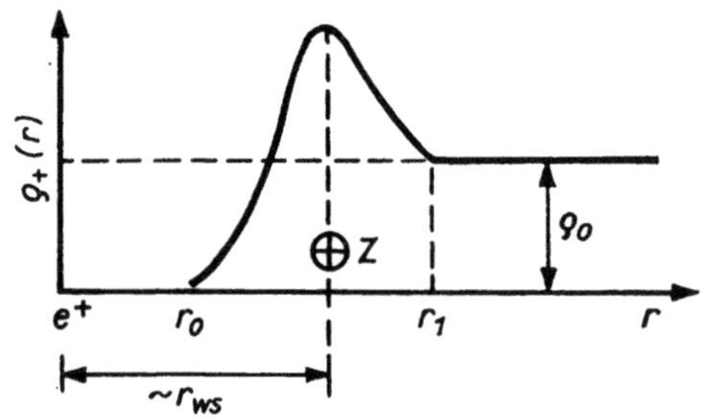

Fig. 1. The assumed distribution of the positive charge around a positron in inhomogeneous jellium. The maximum value of $\varrho_+(r)$ corresponds to the distance of the nearest atom. Detailed description in the text

realistically we use a model in which the positive charge of the background is pushed away to a certain distance from the positron (Fig. 1). The center of the coordinate system is placed on the positron. The distance r_{WS} is of the order of the Wigner-Seitz radius and determines the distance from the positron to the nearest atomic nuclei.

We assume the simplest form of $\varrho_+(r)$ describing qualitatively the shape in Figure 1:

$$\varrho_+(r) = \begin{cases} 0\,, & r \leqq r_0 \\ A(r - r_0)\, e^{-B(r-r_0)}\,, & r_0 < r \leqq r_1 \\ \varrho_0\,, & r > r_1\,. \end{cases} \tag{5}$$

Inserting (5) into Eq. (4) instead of ϱ_0 we get the equation for the total electronic density around a positron.

The function $\varrho_+(r)$ according to (5) contains four adjustable parameters: A, B, r_0 and r_1. Three of them may be determined from the physically obvious conditions:

$$\varrho_+(r = r_1) = \varrho_0\,, \tag{6a}$$

$$\int_{r_0}^{r_1} \varrho_+(r)\, d^3r = \tfrac{4}{3} \pi r_1^3 \varrho_0\,, \tag{6b}$$

$$r_0 + \frac{1}{B} = r_{WS}\,. \tag{6c}$$

3. Results and discussion

The parameter B cannot be determined from physical considerations. It is connected with some discrepancy between our model and physical reality. Its value describes a continuous smearing of jellium charge whereas in fact the positive charge is distributed discontinuously and surrounded by a negative charge.

In Figure 2 we show the annihilation rates obtained for several values of B at three densities: $r_s = 2$, 4 and 6 as a function of the distance r_1. For a given metal r_1 is approximately of the order of r_s. Our model enables to reduce the electronic density on the positron in comparison with the homogeneous jellium model and to enlarge the positron lifetime. This fact is particularly important at higher densities ($2 < r_s \lessapprox 3$) where the annihilation rates obtained by several authors are appreciably higher than the experimental values (see Tab. 1).

Table 1
Annihilation rates at $r_s = 2$ according to several authors. Experimental value for Al ($r_s = 2.07$): $\lambda_{\text{exp}} = 6.3$

Årponen, Pajanne	Lowy, Jackson	Rubaszek, Stachowiak	Gondzik, Stachowiak	Leung et al.
6.8	7	7	7.5	7.6

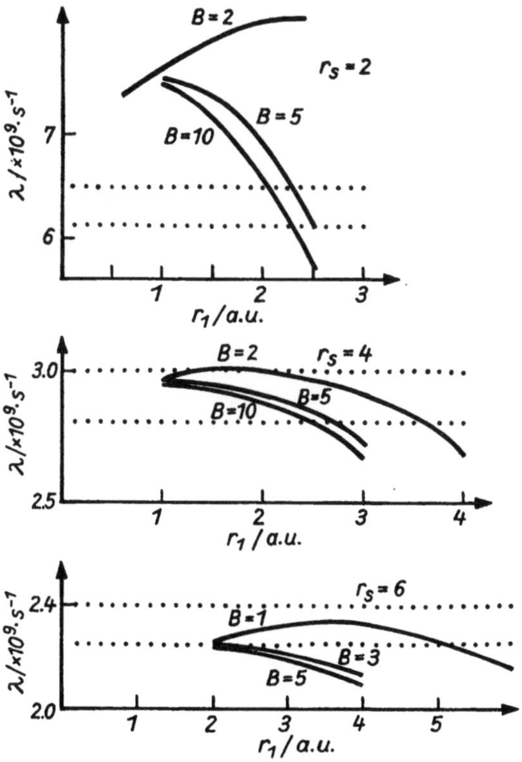

Fig. 2. Annihilation rates for a positron obtained for three values of electron gas density as a function of B and r_1. The dots confine the regions for which the values of λ would correspond to experimental ones with error less than 5 %

Figure 2 shows that the experimental value can be reached by a suitable (though physically reasonable) choice of the parameters.

References

ARPONEN, J., PAJANNE, E.: Ann. Phys. NY **91**, 450 (1975)
BONDERUP, E., ANDERSON, J. U., LOWY, D. N.: Phys. Rev. **B20**, 883 (1979)
CARBOTTE, J. P., SALVADORI, A.: Phys. Rev. **162**, 290 (1967)
CHAKRABORTY, B.: Phys. Rev. **B24**, 7423 (1981)
DANIUK, S., KONTRYM-SZNAJD, G., RUBASZEK, A., STACHOWIAK, H., MAYERS, J., WALTERS, P. A., WEST, R. N.: J. Phys. F, in print
FUJIWARA, K.: J. Phys. Soc. Japan **33**, 1047 (1972)
GONDZIK, J., STACHOWIAK, H.: J. Phys. **C18**, 5399 (1985)
HEDIN, L., LUNDQVIST, B. J.: J. Phys. **C4**, 2064 (1971)
KAHANA, S.: Phys. Rev. **129**, 1622 (1963)
KALLIO, A., PIETILÄINEN, P., LANTTO, L.: Phys. Scr. **25**, 943 (1982)
LEUNG, C. H., STOTT, M. J., ALMBLADH, C. O.: Phys. Lett. **A57**, 26 (1976)
LOWY, D. N., JACKSON, A. D.: Phys. Rev. **B12**, 1689 (1975)
PIETILÄINEN, P., KALLIO, A.: Phys. Rev. **B27**, 224 (1983)
RUBASZEK, A., STACHOWIAK, H.: phys. stat. sol. (b) **124**, 159 (1984)
SJÖLANDER, A., STOTT, M. J.: Solid State Commun. **8**, 1881 (1970)
SORMANN, H., PUFF, W.: Positron Annihilation (eds P. C. JAIN, R. M. SINGRU, K. P. GOPINATHAN), Singapore 1985, p. 161

(Received June 16, 1987)

Authors' address:

M. Sci. J. GONDZIK, Prof. Dr. Sci. H. STACHOWIAK
Institute for Low Temperature and Structure Research
Polish Academy of Sciences,
P.O. Box 937
50-950 Wrocław 2
Poland

Cryst. Res. Technol.	22	1987	12	1515–1518

S. G. Usmar, Y. He, A. Alam

Physics Department, Brookhaven National Laboratory, Upton, New York, USA, and
Department of Physics, University of Bejing, China, and
Department of Physics, University of Ulster, Coleraine BT52 1SA, UK

A Simulation of Vacancy Clustering in Al and its Relevance to Positron Experiments[1])

Preliminary results of the simulation of the vacancy clustering in Al during the quench and subsequent annealing is presented and discussed in view of their relevance to positron experiments.

Vorläufige Ergebnisse der Simulation von Leerstellenclusterung in Al während der Abschreckung und der nachfolgenden Ausheilung werden berichtet und die Relevanz dieser Simulation zu den Positron-Annihilations-Experimenten diskutiert.

1. Introduction

The early stages of vacancy clustering in quenched and irradiated metals (Coleman et al.; Singru et al.) have been the subject of a number of positron studies. These studies have largely been based on the techniques of Doppler-broadening of the electron positron annihilation line and positron lifetime. The Doppler-broadening technique provides a qualitative picture about the clustering process, e.g. whether the migrating species form 3-d clusters at a certain temperature range and whether these 3-d clusters collapse at a later stage to 2-d structures. The positron lifetime can, in principle, provide information about the size and the concentration of the individual clusters present at any one time. However, during the clustering process, the defect population is likely to contain a spectrum of cluster sizes. Consequently, the positron lifetime spectrum is likely to contain a large number of exponential components. In such cases it is almost impossible to resolve the individual lifetime components and their intensities with sufficient accuracy. Therefore, any additional information about the nature of the defects available to assist in the interpretation of the experimental data is always welcome. Such information can be obtained from simple simulations based on a reaction rate model which is widely used in the studies of defects (see e.g. Kiritani; Yoo; Kamigachi et al.). With the simulated defect distribution one can then calculate the various positron annihilation parameters in order to compare with the experimental data. This exercise may also provide insight into various binding and migration energies of the species (which are the parameters used in the simulation process). Recently we have undertaken such studies in relation to some of our experimental data on the recovery behaviour of quenched Al (Alam). Here, we present some preliminary results.

2. The model

First of all we tried to simulate the quenched-in defect population. The starting point was the vacancy concentration at thermal equilibrium at the quenching tem-

[1]) Contributed paper presented during the European Meeting on Positron Studies of Defects, March 23–27, 1987, Wernigerode, G.D.R.

perature. The quenching and bath temperatures were taken to be 850 K and 200 K, respectively (the same as in our experiments). A number of cooling curves were considered based on the estimation of the experimental quenching rate. The quenching process was simulated as isochronal annealing in steps of 5 milliseconds. In each step the input defect population was taken to be the output of the previous step.

The model incorporated the migration of either mono and divacancies or mono-, di- and trivacancies. The interaction between the clusters are given by rate equations such as

$$V_1 + V_{n-1} \underset{G_n^1}{\overset{F_{n-1}^1}{\rightleftharpoons}} V_n,$$

$$V_2 + V_{n-2} \underset{G_n^2}{\overset{F_{n-2}^2}{\rightleftharpoons}} V_n,$$

$$V_3 + V_{n-3} \underset{G_n^3}{\overset{F_{n-3}^3}{\rightleftharpoons}} V_n.$$

Here, V_n is the vacancy cluster composed of n vacancies and F_{n-i}^i and G_{n-i}^i are the rate constants for the association and disassociation processes. The F_j^i's and G_j^i's are given by $F_j^i = f_j \nu_i \exp(-E_{vi}^m/kT)$ and $G_j^i = g_j \nu_i \exp[-(E_{vi}^m + E_{jvi}^b)/kT)]$ where f_j and g_j are geometretical factors (Yoo,) ν_i $(i = 1, 2, 3)$ the vibration frequencies,

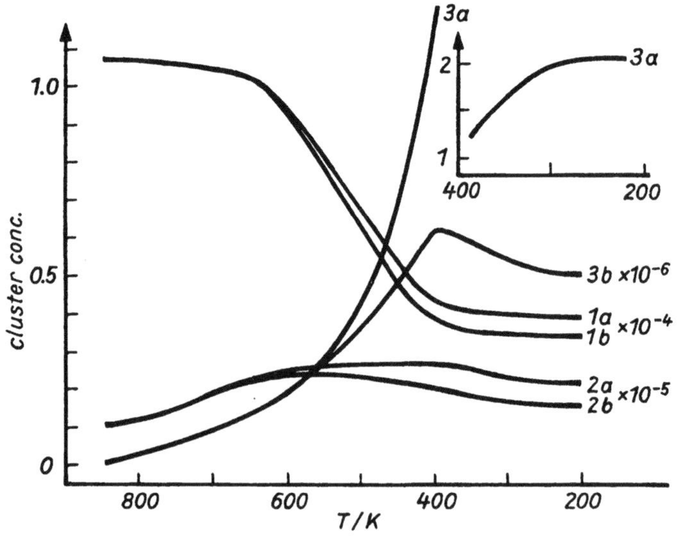

Fig. 1. The concentration profile for mono, di- and trivacancies (1, 2, 3) during the quench: mobile (a) mono- and divacancies, (b) mono-, di- and trivacancies

E_{vi}^m the migration energy for the defect and E_{jvi}^b is the binding energy of the defect to the cluster. f_j, g_j, and E_{jvi}^b are estimated according to the prescription given in Yoo. We considered clusters of upto 30 vacancies and in our initial simulations the loss of vacancies into sinks was ignored. Some initial results for the quenching process is given in Figure 1 which shows the changes in the concentration of mono-, di- and trivacancies during the quench. We find that the quenched-in population of mono-

and di-vacancies in only marginally dependent on whether the trivacancy is mobile. However, when the trivacancies are mobile their concentration reduces by a factor of ~ 4 at the end of the quench. Also, by varying the maximum cluster size between 10 to 30 vacancies the concentration of the mono-, di- or trivacancies does not undergo appreciable change during the quench. The average positron lifetime, calculated (using lifetime values given by HAUTOJÄRVI et al.) from the quenched-in population which incorporates the migration of trivacancies, matches reasonably well with our measured average lifetime.

With these quenched-in defect population we simulated isothermal annealing at a temperature of 215 K (where the clustering peak is observed in the experimental

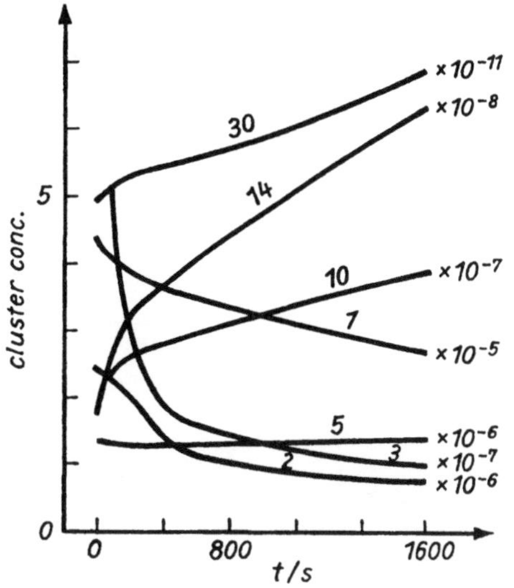

Fig. 2. The concentration profile of vacancy clusters during isothermal annealing at 215 K (mobile mono-, di- and trivacancies). The numbers along each line indicate the number of vacancies in a cluster

data). Figure 2 shows some of the initial results. Here, again we found similar dependency of the mono-, di- and trivacancy concentrations on the mobility of trivacancies (not shown in the figure). Also, the average lifetime was calculated after an anneal of 2000 s and fairly good agreement was reached with the measured average positron lifetime data after an anneal of 2000 s at this temperature. To obtain a more realistic picture of the simulated defect distribution, obviously, the loss of vacancies to sinks and the creation of planar defects have to be taken into consideration. These will be introduced in our future efforts. However, even at this stage we feel that the information contents of positron annihilation experiments can be largely enhanced by such studies.

References

ALAM, A.: unpublished results
COLEMAN, P. G., SHARMA, S. C., DIANA, L. M.: Positron Annihilation, Amsterdam 1982
HAUJOTÄRVI, P., HEINIO, J., MANNINEN, M., NIEMINEN, R. M.: Phil. Mag. 35, 973 (1977)
KAMIGACHI, N., HASHIMOTO, E., DEGUCHI, Y., KINO, T.: J. Phys. Soc. Jap. 51, 583 (1982)

Kiritani, M.: J. Phys. Soc. Jap. **35**, 95 (1973)
Singru, R. M., Jain, P. C., Gopinatan, K. P.: Positron Annihilation, Singapore 1986
Yoo, M. H.: J. Phys. **F11**, L65 (1981)

(Received June 16, 1987)

Authors' addresses:

Dr. S. G. Usmar
Physics Department
Brookhaven National Lab.
Upton New York, USA

Dr. Y. He
Physics Department
University of Bejing
China

Dr. A. Alam
Department of Physics
University of Ulster
Coleraine BT52 1SA, UK

Cryst. Res. Technol.	22	1987	12	1519–1522

P. Beasley, A. Alam, R. N. West

School of Maths & Physics, University of East Anglia, Norwich, U.K., and
Department of Physics, University of Ulster, Coleraine, U.K.

The Interaction of Ni and Au with HCl during Quenching[1])

During annealing of Ni and Au, previously quenched into HCl at 203 K, annealing stages that cannot easily be explained in terns of the migration of open volume defects are observed. The contamination of the hot samples with hydrogen released from the quenching fluid is suggested as the origin of these additional annealing stages.

Während der Ausheilung von abgeschreckten Ni und Au (in HCl, 203 K) haben wir Ausheilungsstufen beobachtet, die nicht als Auswanderung von offenen Defektarten erklärbar sind. Eine Kontamination der heißen Probe mit Wasserstoff, der von HCl freigelassen wird, wird als die Ursache dieser Ausheilungsstufen vorgeschlagen.

1. Introduction

In recent years the positron annihilation technique has proved to be an effective tool in the study of 3-dimensional vacancy clusters. Because, for small clusters, there is a strong dependence of the annihilation parameters on the cluster size (Nieminen, Manninen) the technique is particularly useful for study of the nucleation and early stages of cluster growth following deformation, irradiation, or quenching of metals and alloys (see for example the various reports of Coleman et al.; Singru et al.). In quenching studies one tries to quench in the maximum possible number of vacancies (created at thermal equilibrium at the initial high temperature) and to minimize possible clustering or loss of these during the quench. In this, the choice of quenching medium is vital. However, in this choice, care has also to be taken in order to avoid or to minimize the interaction of the specimens with the quenching medium. Over the past few years, concentrated HCl has proved a useful medium, particularly for Al and dilute alloys of Al (Alam, West; Hood, Schultz). The advantage of HCl is that, apart from its thermal capacity, it can be cooled down to the order of 200 K, a temperature that is sufficiently low to prevent the migration of vacancies in most metals. In some earlier work we compared the recovery of Al and dilute Al-based alloy specimens quenched variously into HCl, methanol, and liquid N_2. The results suggested no significant reaction of the specimens with the HCl. In this note we report the results of a similar study of Ni and Au but one that realizes rather different conclusions.

2. Experimental details

The samples studied were polycrystalline Ni (99.998%) and Au (5N) obtained from commercial suppliers. The specimens were pre-heated in a pure nitrogen atmosphere (5 N)

[1]) Contributed paper presented during the European Meeting on Positron Studies of Defects, March 23–27, 1897, Wernigerode, G.D.R.

at 0.9 of the melting temperature (T_m) and then quenched, either into HCl, precooled to 203 K, or into liquid nitrogen. Immediately following a quench into HCl, the samples were washed in methonl at 200 K to remove the acid and then dried on a pre-cooled absorbent paper placed on a copper table at 77 K. The same cold table and component precooling technique was used in the subsequent and rapidly implemented source-sample construction in which two identically treated specimens were sandwiched around a 15 μCi ^{22}Na source sealed between two 2 μm foils. Apart from the methonal wash the N_2 quenched specimens were similarly treated. By this means the total time spent by the specimens above 150 K was kept to a minimum (60 s maximum). The annealing behaviour was studied through measurements of the Doppler broadening lineshape S-parameter. Energy distributions were recorded at 2000 s intervals while the specimens were annealed, in-situ, during continuous heating at a linear rate of ~ 6 Kh^{-1} (ALAM, WEST).

3. Results and discussion

The annealing curve for HCl-quenched Ni (Fig. 1a) shows interesting features. The initial value of S-parameter is consistent with a significant degree of trapping in quenched-in defects — presumably vacancies and small vacancy clusters produced

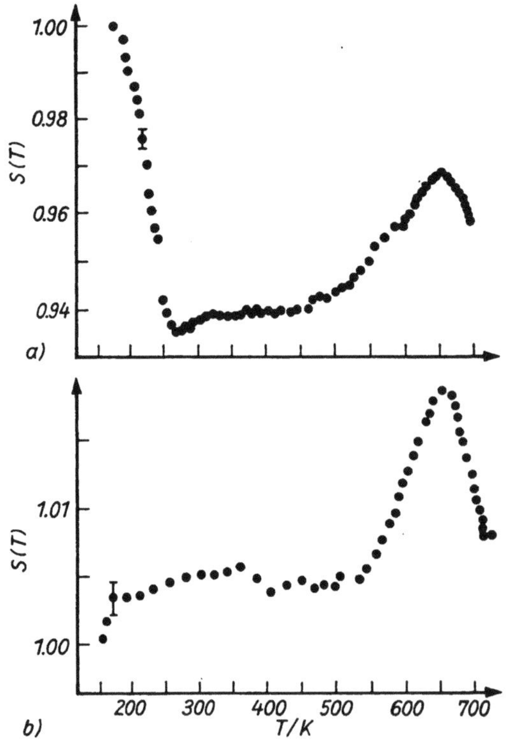

Fig. 1. The annealing behaviour of Ni quenched into (a) an HCl Bath at 203 K and (b) a liquid nitrogen bath. The initial S values are arbitrarily normalised to 1

during the quench. The large drop in $S(T)$ at ~ 150 K indicates a corresponding decrease in this trapping which means that either the vacancy traps are lost to sinks during migration or that their effectiveness as positron traps is inhibited by some mechanism. However, we know from annealing studies of deformed specimens (DLUBEK et al.; BEASLEY et al.) that vacancies in Ni do not migrate below room temperature.

Migration of vacancies at this temperature would also seem to be incompatible with the eventual rise in $S(T)$ that occurs at around 500 K (Fig. 1a) for such an increase is invariably associated with vacancies moving and the resulting formation of 3-d clusters. All these factors lead us to conclude that the migrating species at 150 K are something other than vacancies.

It has been suggested (DLUBEK) that during the quench in the HCl the acid dissociates around the hot specimen and that the hydrogen thus released diffuses into the bulk. Following others (COLLINS, SCHUMANN; HOHLER, KRONMULLER; HANSEN, RAJAINMAKI) we associate the stage at 150 to 250 K with the migration of this hydrogen to the open-volume defects. The consequent inhibition of positron trapping can then explain the observed fall in the S-parameter. There is further evidence to support this conjecture in the curves for recovery following the quench into liquid N_2.

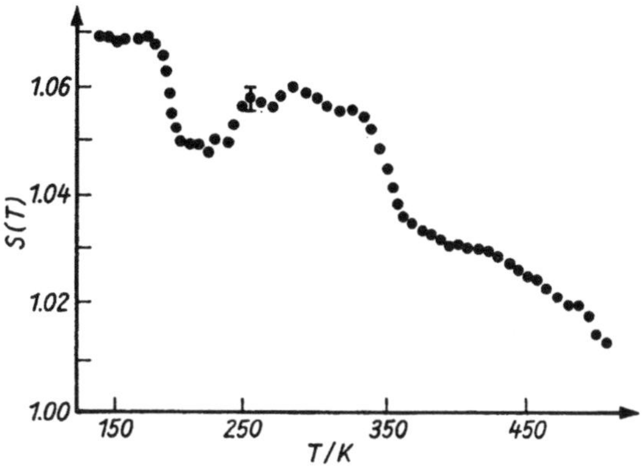

Fig. 2. Recovery behaviour of Au quenched into an HCl bath at 203 K. The initial S value has been arbitrarily normalised to 1

Here (Fig. 1b), there is no recovery stage at 150 K, but the higher temperature stage at ~ 600 K already seen in Figure 1a is even more evident. This latter fact is significant. Of the two quenches, that in the liquid N_2 is bound to have been the less efficient and to have realized fewer vacancies. Yet in Figure 1b the vacancy clustering peak at 600 K is markedly larger. This must mean that in the annealing represented by Figure 1a only a fraction of the vacancies in the as-quenched specimen survice as individual or agglomorate positron traps to 600 K.

When Au is quenched in HCl a similar although somewhat smaller effect is oberved in its subsequent annealing. Here (Figure 2), we observe a fall in $S(T)$ between 185 and 205 K and a subsequent rise at around 250 K (for a fuller, albeit tentative, interpretation of the individual stages the reader is referred to BEASLEY et al.). The results of perturbed angular correlation (PAC) studies by DEICHER et al. (1981) of irradiated and quenched (under almost identical conditions to those of our experiments) Au have been used to suggest that a small defect cluster in Au migrates at ~ 195 K and the monovacancy at ~ 250 K. However, although those quoted temperatures roughly coincide with those of the various stages apparent in Figure 2, the dip in $S(T)$ at ~ 195 K is, in those terms, not too easily explained.

The curve of Figure 2 is very similar in its essential aspects to that of Figure 1a. This leads us to suggest a similar set of underlying mechanisms at least for the initial fall in $S(T)$. That and the less dramatic effect (compared to that in the annealing of Ni) suggests that Au also reacts with the HCl bath but to a lesser extent.

We would like to thank the SERC, Swindon for their support of this work and Dr. G. Dlubek for invaluable discussions. Thanks are also due to Miss Karen Fleming for her help in the preparation of this manuscript.

References

ALAM, A. and WEST, R. N.: J. Phys. **F12** 389 (1982)
BEASLEY, P., WEST, R. N., ALAM, A.: Mat. Sc Forum **15**, 169, 175 (1987)
BSASLEY, P., ALAM, A., USMAR, S. G., COLEMAN, P. G., WEST, R. N.: in: Positron Annihilation, eds R. M. SINGRU et al., Singapore 1986
COLEMAN, P. G., SHARMA, S. C., DIANA, L. M.: Positron Annihilation, Amsterdam 1982
COLLINS, G. S., SCHUHMANN, R. B.: Phys. Rev. **B34**, 502 (1986)
DEICHER, M., RECKNAGEL, E., WICHERT, Th.: Rad. Effects **54**, 155 (1981)
DLUBEK, G.: 1985 (private communication)
DLUBEK, G., GRÜMMER, O., MEYENDORF, N.: Sol. State Comm. **27**, 1219 (1978)
HANSEN, H. E., RAJAINMAKI, H.: in: Positron Annihilation, eds R. M. SINGRU et al., Singapore 1986
HOHLER, B., KRONMÜLLER, H.: Phil. Mag. **A43**, 1189 (1981)
HOOD, G. M., SCHULTZ, R. J.: in: Positron Annihilation, eds R. M. SINGRU et al., Singapore 1986, p. 506
NIEMINEN, R. M., MANNINEN, M.: in: Positron Studies of Solids, ed. P. HAUTOJÄRVI, 1979 (Topics in Current Physics Vol. 12)
SINGRU, R. M., JAIN, P. C., GOPINATHAN, K. P.: Positron Annihilation, Singapore 1986

(Received June 16, 1987)

Authors' addresses:

Mr. P. BEASLEY
School of Maths & Physics
University of East Anglia
Norwich NR4 7TJ, UK

Dr. A. ALAM
Department of Physics
University of Ulster
Coleraine BT52 1SA, UK

Prof. R. N. WEST
School of Maths & Physics
University of East Anglia
Norwich NR4 7TJ, UK

Cryst. Res. Technol.	22	1987	12	1523–1527

St. CHABIK

Institute of Physics, Pedagogical University, Opole, Poland

Vacancy Formation Enthalpy in AgZn Alloys[1])

Effective vacancy formation enthalpy (H_v) for AgZn alloy was determined by means of positron annihilation method. It was found that H_v decreases linearly with increasing Zn content for the α-phase. For the ξ and β'-phase the H_v does not change and is equal to 0.53–0.55 eV.

Mit Hilfe des Positronen-Annihilationsverfahrens wurde die effektive Enthalpie der Leerstellenbildung (H_v) für AgZn-Legierungen bestimmt. Die Enthalpie wächst linear mit wachsender Zn-Konzentration für die α-Phase. – Für die Phasen ξ und β' ist die Enthalpie konstant und beträgt 0,53–0,55 eV.

Previously CHABIK and ROZENFELD reported on the results of measurements of the effective vacancy formation enthalpy for AgZn alloys containing 29.2 and 50 at.% Zn, suggesting that similarly as for CuZn system the vacancy formation enthalpy in AgZn alloys decreases with increasing Zn content. For chocking this suggestion the measurements were repeated for AgZn alloy containing 14.6, 21.9 (α-phase), 35.5 ($\alpha + \xi$), 43.7 and 47 (ξ-phase) at.% Zn. The effective vacancy formation enthalpy (H_v) has been determined from equilibrium experiments by measuring the temperature dependence of the peak counting rate of positron annihilation quanta. The experimental results were elaborated according to a trapping model, according to which the temperature of the normalized peak counting rate can be represented by formuala

$$F(T) = \frac{A \cdot F_v \exp\left(-\frac{H_v}{kT}\right)}{1 + A \exp\left(-\frac{H_v}{kT}\right)} + F_0 (1 + \gamma \cdot \Delta T) , \quad (1)$$

where F_0 and F_v denote the values of the $F(T)$ parameter in the case when positrons are not trapped at vacancies, and in the case when almost all the positrons annihilate from trapped state, respectively; $A = \mu\tau$, μ – positron trapping rate at vacancies, τ – mean value of the positrons lifetime, γ – annihilation coefficient for thermal expansion.

The results of measurements for five investigated alloys of different composition are shown in Figures 1–5. The solid lines in these figures represent the curve calculated according to eq. (1), using for A and H_v the values collected in Table 1. The fourth column contains the value of the coefficient γ for samples showing a linear change in the $F(T)$ parameter in the sub-vacancy region.

For alloy containing 47 at.% Zn in the vicinity of the β-ξ- phase transition the experimental points deviate from the S-shaped curve. The temperatures of the phase transformation for alloys containing 43.7 and 47 at.% of Zn were determined from the X-ray measurements in the heating cycle equal to 270 ± 0.5 °C and 273.5 ± 0.5 °C for 43.7 and 47 at.% Zn alloy, respectively. The DTA technique gives for 43.7 at.% Zn

[1]) Contributed paper presented during the European Meeting on P Positron Studies of Defects, March 23–27, 1987, Wernigerode, G.D.R.

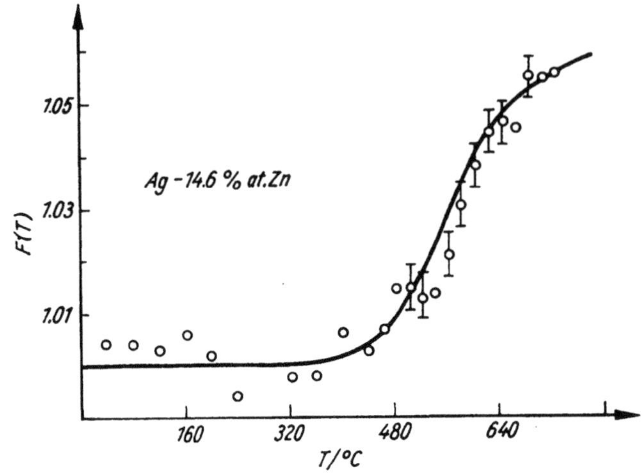

Fig. 1.

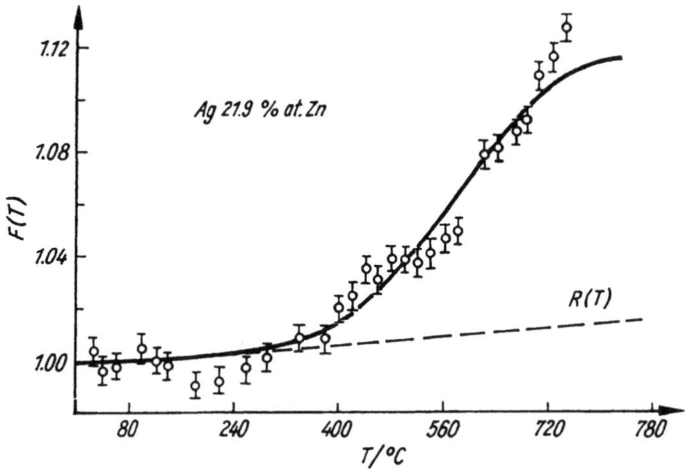

Fig. 2.

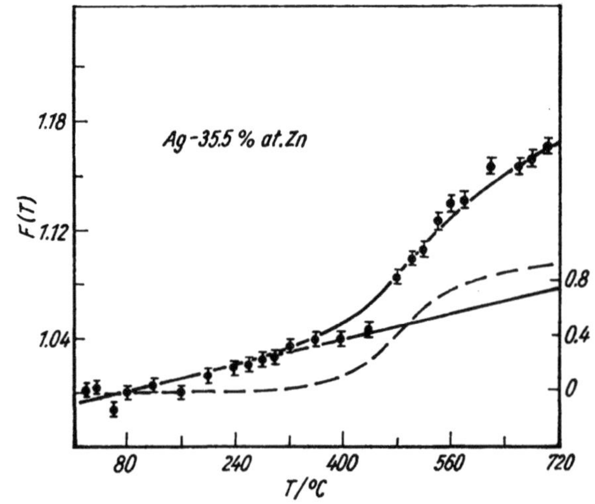

Fig. 3.

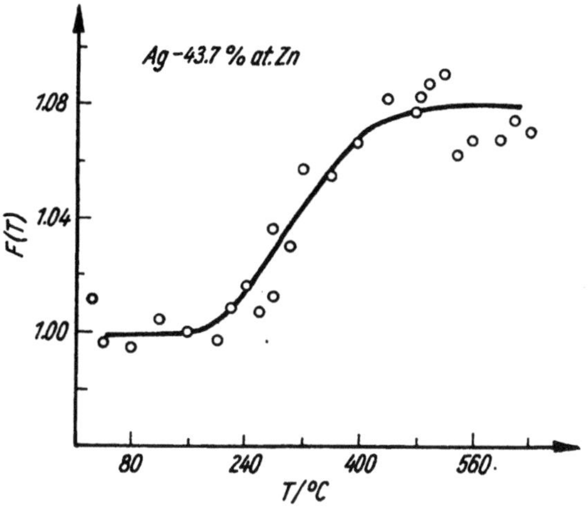

Fig. 4.

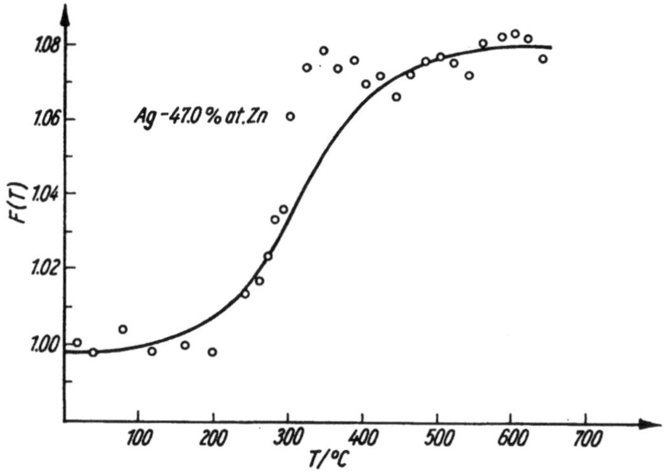

Fig. 5.

Fig. 1 – 5. The temperature dependence of $F(T)$ parameter calculated from formula (1) (solid line) and experimental points for different AgZn composition

Table 1
Parameters representing the results of the positron annihilation on AgZn alloys

Zn-content percent	H_v eV	$A \times 10^5$	$\times 10^4$ K^{-1}
14.6	1.05 ± 0.05	5 ± 0.6	
21.9	1.00 ± 0.03	1.3 ± 0.6	0.22 ± 0.04
35.5	0.91 ± 0.05	8.9 ± 0.4	1.20 ± 0.10
43.7	0.55 ± 0.03	0.8 ± 0.4	
47.0	0.53 ± 0.03	0.4 ± 0.2	

alloy 277 ± 1 °C and 255 ± 1 °C in the heating and cooling runs, respectively, the corresponding temperatures for 47 at.% Zn alloys being 280 and 261 °C. It is noteworthy that the results obtained in the DTA measurements are independent on the heating (cooling) rates. The high temperature studies of zinc containing alloys are inherently connected with dezincification phenomenon changing the energetical conditions and the chemical composition of investigated materials. Taking this into account as well as the results of DTA measurements, the $F(T)$ parameter for 43.7 and 47 at.% Zn alloys were performed in two heating and cooling runs. The results for 47 at.% Zn alloy are shown in Figure 6. As it is seen, the course of the $F(T)$ curve in the heating run is relatively smooth, whereas in the cooling run a jump in the $F(T)$ value occurs in the temperature interval 260—270 °C, i.e., during the phase transition. This indicates a difference in the kinetics of the $\beta \to \xi$ and $\xi \to \beta$ phase transition.

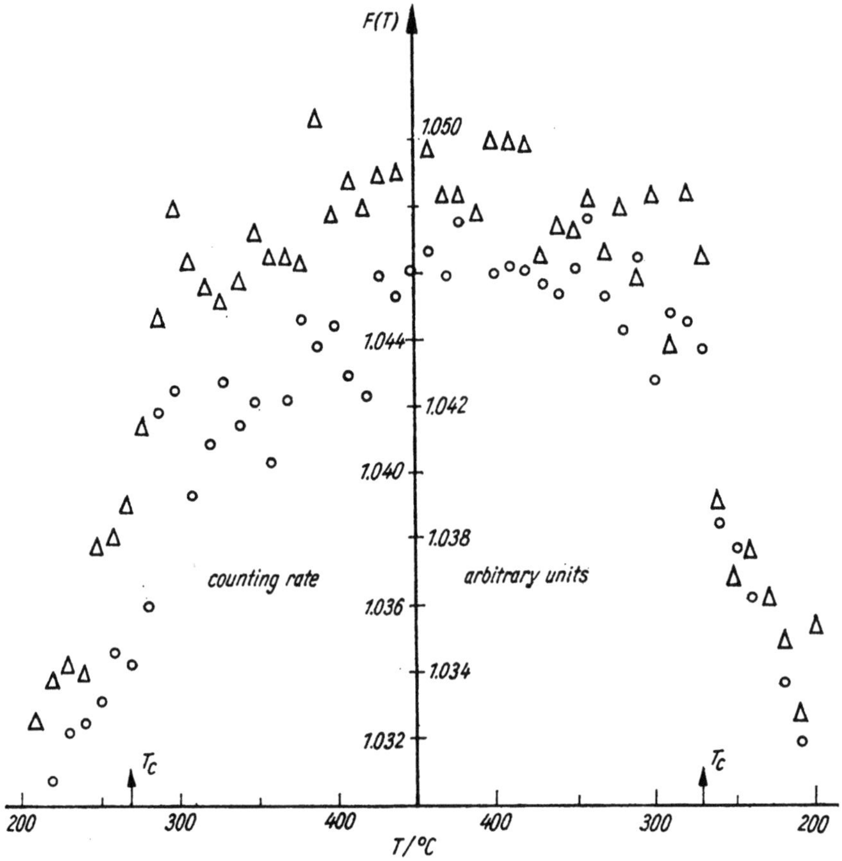

Fig. 6. The temperature dependence of $F(T)$ for Ag 47 at.% Zn for first (○) and second (△) cycles. T_c denote $\beta \to \xi$ transition

This is difficult to decide whether the effect is connected with the dezincification of the surface layers or with the specific nature of the transition, although the DTA measurements — with results not so sensitive to the surface effects — tend to support this last supposition. It is worth to note that the $F(T)$ courses are in both the cycles the same, i.e., are insensitive to the dezincification, although the $F(T)$ values at corresponding temperatures are in the second cycle larger. The last effect is most probably due to the increase in the defectiveness of the sample due to the thermocycling.

The results of the previous paper (CHABIK, ROZENFELD) and the present study are collected in Figure 7 as the concentration dependence of the effective vacancy formation enthalpy in AgZn alloys. The values of H_v for pure silver were taken from

TRIFTSHÄUSER, MACGERVEY, from LICHTENBERG et al. and JEAN et al. It can be concluded that:
— the vacancy formation enthalpy decreases linearly with increasing Zn content:

$$H_{v\,AgZn} = H_{v\,Ag} - 0.023\, c_{Zn}[eV]\,,$$

where c_{Zn} denotes the zinc content in at.%,
— the vacancy formation enthalpy in the ξ-phase does not depend on the zinc content equal to 0.53—0.55 eV,
— the annihilation mechanism close to T_c depends on the direction of the phase change.

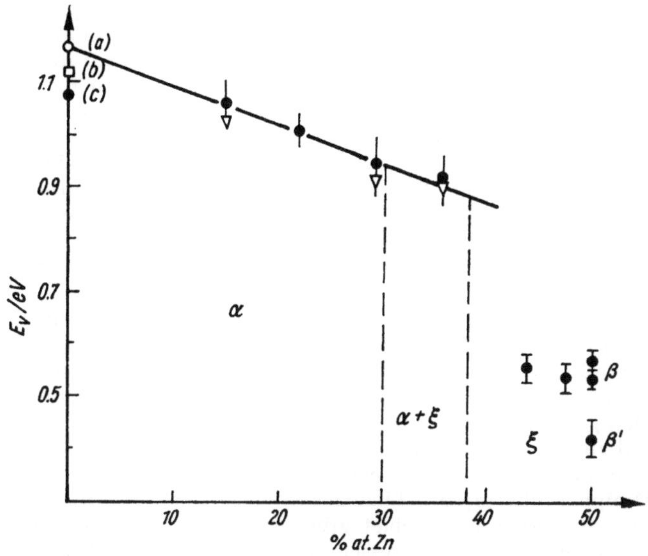

Fig. 7. Dependence of H_V for AgZn. Value of H_V for pure Ag taken from TRIFTSHÄUSER, MACGERVEY, from LICHTENBERG et al. and JEAN et al.

The concentration dependence of the effective vacancy formation enthalpy in AgZn alloy is similar that to for CuZn alloys.

This work has been supported by scientific program CPBP 01.06.

References

CHABIK, St., ROZENFELD, B.: Positron Annihilation, Amsterdam 1982, p. 547
JEAN, Y. C., LYNN, K. G., DICMAN, I. E.: Phys. Rev. **B21**, 2655 (1980)
LICHTENBERG, P. C., GINGERICH, R. R., CAMBELL, I. L.: Phys. Lett. (a) **41**, 305 (1972)
TRIFTSHÄUSER, W., MAC ERVEY, J. D.: Appl. Phys. **6**, 177 (1979)

(Received June 16, 1987)

Author's address:

St. CHABIK
Institute of Physics, Pedagogical University
Oleska 8
45052 Opole
Poland

Book Review

Ch. M. STARKS (ed.)

Phase-Transfer Catalysis

ACS Symposium Series, vol. 326

American Chemical Society, Washington 1987, 195 + ix, 44.95 US $, ISBN 0-8412-1007-1

Auch unter seriösen Verlegern ist es mittlerweise en vogue, die Hauptvorträge von Tagungen zu bündeln, als Buch zu kaschieren und damit die Entwicklungsphase auf mindestens die Hälfte zu drücken, indem ebenfalls die formale Arbeit auf die Autoren durchgegeben wird. Die Symposienreihe der amerikanischen chemischen Gesellschaft macht daraus wenigstens keinen Hehl, und man weiß von Anfang an, woran man ist: Vol. 326, Fortschrittsbericht eines gleichnamigen Symposiums auf dem Jahrestreffen der ACS 1985 in Chicago, bietet er wohlaufbereitet (die Autoren hatten ein knappes Jahr Zeit) auf knapp 200 Seiten gut illustriert und mit Tabellen, Abbildungen, Reaktionsgleichungen und Formelschemata versehen, das Material von 14 Hauptvorträgen zum Thema Phasentransferkatalyse. Davon die Hälfte aus namhaften US Chemiefirmen, die andere Hälfte aus Universitäten und technischen Instituten, 2 aus Kanada, 1 aus Italien, mit summiert 450 aktuellen Referenzen.

Phasentransferreaktionen sind chemische Molekülreaktionen in (flüssigen) Mehrphasensystemen. Sie bedürfen spezieller Katalysatoren, die den Stoffwandlungsprozeß zwischen den Phasen bewerkstelligen. Deshalb verlaufen sie sehr selektiv. In den 1986er Listen bieten marktführende Feinchemikalienfirmen über 200 Phasen-Transfer-Katalysatoren ("PTC") an. Seit den grundlegenden Arbeiten Ende der 70er Jahre von M. MAKOZA und A. BRÄNDSTRÖM sind zu den ursprünglichen quaternären Ammonium- und Phosphoniumsalzen inzwischen Kronenether, Chelatreagenzien und unlösliche polymergebundene Katalysatoren („Dreiphasenkatalyse") hinzugekommen. Nachteil: Sie sind alle ziemlich teuer. Für das rasch expandierende Gebiet gibt es bereits Monographien, und themenrelevante Tagungen sprechen viele Interessenten an. Siehe das vorliegende Buch. Seine behandelten Schwerpunkte lassen sich grob in A. Theorie und Mechanismen, B. Katalysatorentwicklung, C. Anwendung in der Polymerchemie gliedern. Aufgefallen sind mir die offenbar aussichtsreichen Versuche, in einer Art HOFMANN-Reaktion PTC zur Aufbrechung der Polysulfidvernetzungen in Altgummi zu nutzen, dabei die Hauptkette intakt zu lassen und die Struktur neuen Gummis anzustreben (Goodrich). Für die Carbonylierung und Carboxyalkylierung mittels organometallischer Anionen eignet sich anionisch aktiviertes Al_2O_3 (immobilisierte Polyethylenglykole) am besten (Texaco). Sogar freie Radikal-Polymerisationen lassen sich PT-katalysieren ($K_2S_2O_8$/Cyclodextrin bzw. Aliquat 336 = Tricapryltrimethylammoniumchlorid) (3 M). Man sucht verstärkt nach PTC, die sich bei hohen Temperaturen einsetzen lassen, z.B. gewisse Polyethylenglykole (General Electric). Chirale PTC schließlich führen z.B. zu asymmetrischer Alkylierung (Merck & Co.). Ein interessantes Symposium und gar kein schlechtes Buch.

E. HOYER

Cryst. Res. Technol.	22	1987	12	1529–1532

C. S. Sundar, A. Bharathi, K. P. Gopinathan

Materials Science Laboratory, Indira Gandhi Centre for Atomic Research, Kalpakkam, India

Study of Clustering of Vacancies in Cold-worked Nickel by Positron Annihilation Spectroscopy[1)]

The clustering of vacancies occurring during the isochronal and isothermal annealing treatments of a 40% cold-worked nickel has been investigated through positron lifetime measurements. The observed variation in lifetime has been quantitatively understood by modelling the defect reactions amongst vacancies leading to the formation of vacancy clusters.

1. Introduction

Positron annihilation spectroscopy (PAS) is sensitive to probe submicroscopic vacancy clusters. The formation of small vacancy clusters during the isochronal annealing treatments of several metals have been investigated by PAS (Petersen). Dlubek et al. have investigated the isochronal annealing behaviour of cold-worked nickel and have observed that vacancies cluster to form three dimensional microvoids in the temperature range of 100 °C to 400 °C. In the present experiments on cold-worked nickel, in addition to isochronal annealing, isothermal annealing experiments have also been carried out in order to investigate the kinetics of vacancy clustering. Further, calculations have been made to understand the observed variation in lifetime in terms of the evolution of average cluster size as obtained from the reaction-rate theory model for clustering of vacancies in nickel (Yoo).

2. Experimental details

Nickel foils of 99.98% purity (obtained from Lieco Industries, USA) were cold-rolled at room temperature to an extent of 40% increase in length. These were subjected to isochronal annealing treatments in the temperature range of 25 °C to 580 °C. Positron lifetime measurements were carried out at room temperature using a fast-slow coincidence spectrometer having a resolution of 280 ps FWHM. Isothermal annealing treatments on the cold-worked sample were carried out at the aging temperature of 285 °C and 360 °C for various periods of time upto a total of 60 minutes. The lifetime spectra were analysed using the programme of Kirkegaard and Eldrup.

3. Results and discussion

The variation in lifetime components τ_1, τ_2 and the intensity I_2 as a function of isochronal annealing temperature is shown in Figure 1. In the as-deformed sample, a single lifetime component with a value of 170 ps is observed and this is attributed

[1)] Contributed paper presented during the European Meeting on Positron Studies of Defects, March 23–27, 1987, Wernigerode, G.D.R.

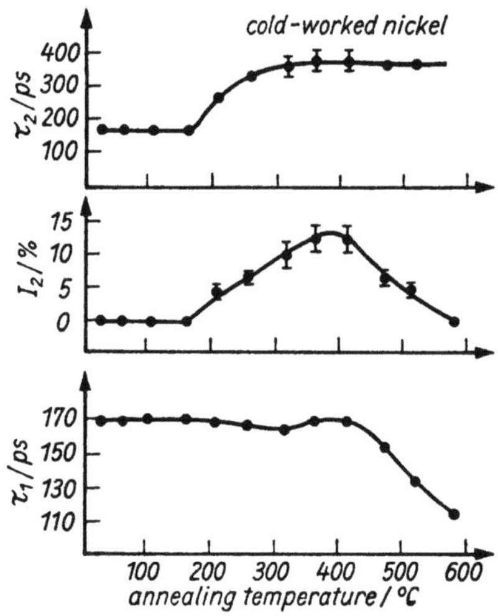

Fig. 1. Variation of positron lifetime parameters τ_1, τ_2 and I_2 as a function of isochronal annealing temperature

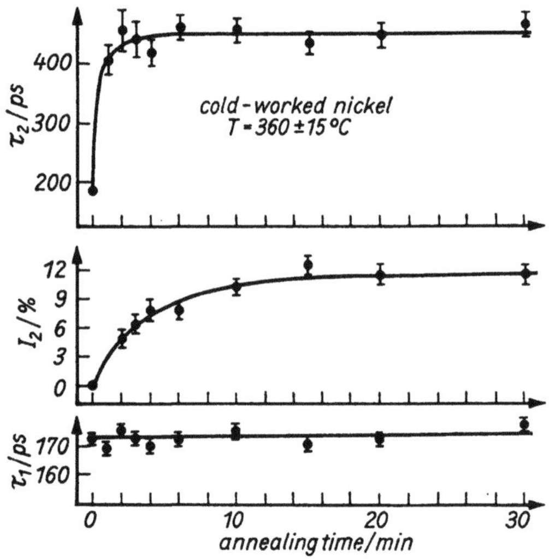

Fig. 2. Variation of τ_1, τ_2 and I_2 as a function of isothermal annealing time at $T = 360$ °C

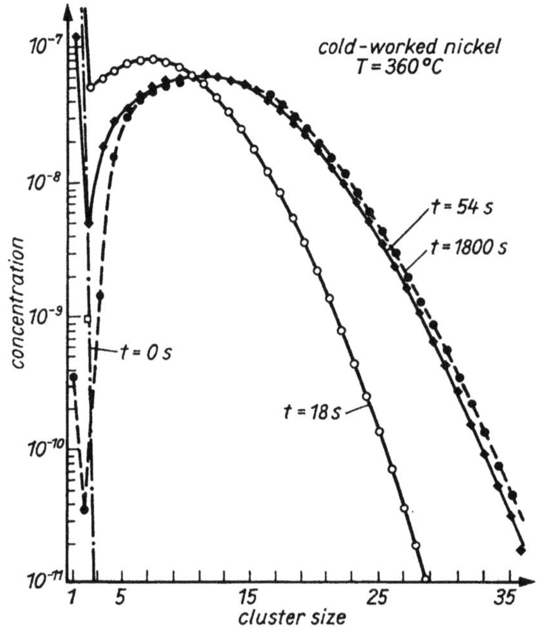

Fig. 3. Evolution of cluster distribution with time at $T = 360$ °C

to complete positron trapping at vacancies and dislocations produced during the cold-working. At temperatures beyond 160 °C, the lifetime spectra develop a second component with the lifetime τ_2 and the intensity I_2 increasing with temperature upto 400 °C. This is attributed to positron trapping at vacancy clusters formed due to the agglomeration of migrating vacancies. The observed decrease in I_2 beyond 400 °C is understood in terms of the dissociation of vacancy clusters. The pertinent result of the present isochronal annealing measurements is the clustering of vacancies observed in the temperature range of 160 to 400 °C and this is similar to the earlier observations of DLUBEK et al.

Positron lifetime measurements as a function of isothermal annealing time at the aging temperatures of 285 °C and 360 °C were carried out in order to investigate the kinetics of clustering of vacancies. Results of the isothermal annealing measurements at 360 °C are shown in Figure 2. The observed increase of τ_2 from a value of 170 ps, characteristic of monovacancies, represents the evolution of vacancy clusters from the monovacancies present in the as-deformed sample. The corresponding increase in I_2 is due to an increase in the positron trapping rate per defect as the vacancy clusters grow with annealing time.

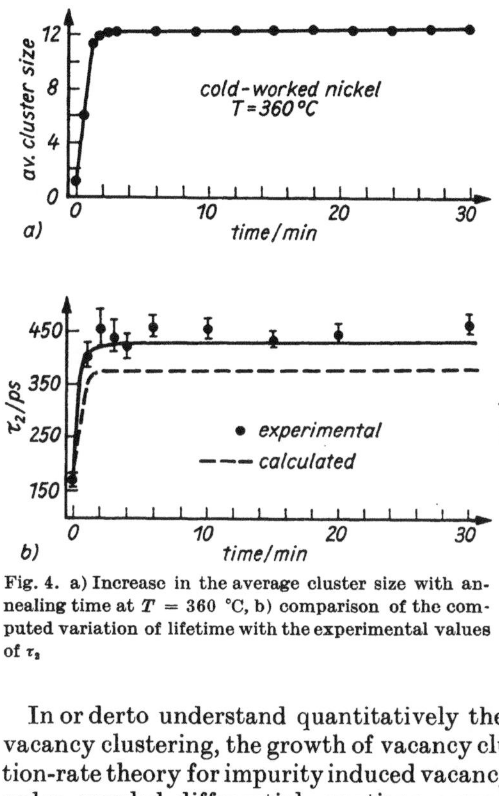

Fig. 4. a) Increase in the average cluster size with annealing time at $T = 360$ °C, b) comparison of the computed variation of lifetime with the experimental values of τ_2

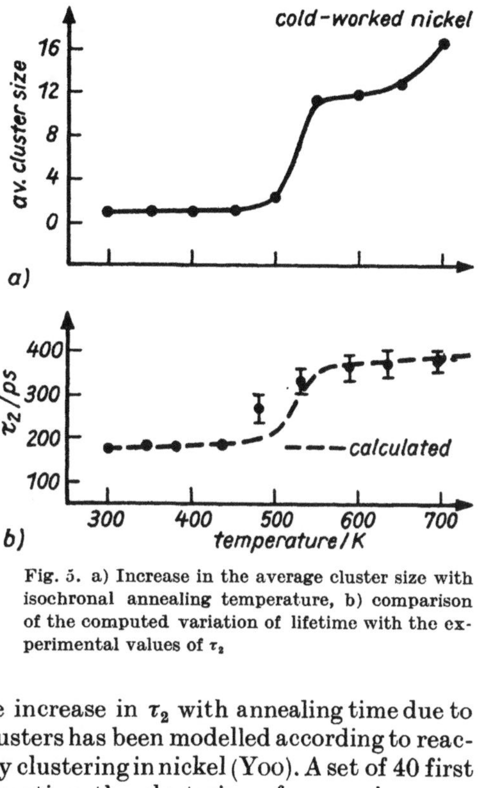

Fig. 5. a) Increase in the average cluster size with isochronal annealing temperature, b) comparison of the computed variation of lifetime with the experimental values of τ_2

In or derto understand quantitatively the increase in τ_2 with annealing time due to vacancy clustering, the growth of vacancy clusters has been modelled according to reaction-rate theory for impurity induced vacancy clustering in nickel (YOO). A set of 40 first order coupled differential equations, representing the clustering of vacancies, were solved numerically on a computer to obtain the evolution of cluster distribution with annealing time at various aging temperatures. The evolution of cluster distribution at $T = 360$ °C is shown in Figure 3 and the increase in the average cluster size with annealing time is shown in Figure 4a. This coupled with the known values of positron lifetimes at vacancy clusters in nickel (PUSKA and NIEMINEN) is used to construct the variation of lifetime with annealing time. The results are shown in Figure 4b along with the experimental values of τ_2 vs annealing time. The evolution of average cluster size as a function of isochronal annealing temperature as obtained from the solution of rate equations is shown in Figure 5a. A plot of the computed variation of lifetime vs annealing temperature along with the experimental values of τ_2 is shown in Figure 5b.

To conclude, the trends in the observed variation of τ_2 with annealing temperature and time are well reproduced by the reaction-rate theory model for vacancy clustering.

References

DLUBEK, G., BRÜMMER, O., MEYENDORF, N., HAUTOJÄRVI, P., VEHANEN, A., YLI-KAUPPILA, J.: J. Phys. F: **9**, 1961 (1979)

KIRKEGAARD, P., ELDRUP, M.: Comp. Phys. Commun. **7**, 401 (1974)

PETERSEN, K.: in: Positron Solid State Physics, Ed.: W. BRANDT and A. DUPASQUIER, Amsterdam 1983, p. 298

PUSKA, M. J., NIEMINEN, R. M.: J. Phys. F: **13**, 333 (1983)

YOO, M. H.: J. Phys. F: **11**, L65 (1981)

(Received June 16, 1987)

Authors' address:

C. S. SUNDAR, A. BHARATHI, K. P. GOPINATHAN
Materials Science Laboratory,
Indira Gandhi Centre for Atomic Research
Kalpakkam — 603102, India

Cryst. Res. Technol.	22	1987	12	1533–1535

J. DEL RÍO, N. DE DIEGO, C. HIDALGO

Facultad de Ciencias Físicas, Departamento de Física de la Materia Condensada
Universidad Complutense, E-28040 Madrid, Spain

Positron Trapping in Deformed Gd[1])

Positron lifetime and Doppler broadening measurements have been used to study the annealing behaviour of deformed Gd between 290 and 700 K. The observed recovery stage at about 370 K is attributed to dislocations annealing.

Positronenlebensdauer- und Annihilationslinienbreite-Messungen wurden angewandt, um das Verhalten von verformtem Gd im Temperaturbereich zwischen 290 und 700 K zu untersuchen. Bei etwa 370 K heilen Defekte aus, die durch Verformung erzeugt wurden. Diese Defekte werden als Versetzungen identifiziert, die mit Fremdatomen dekoriert sind.

1. Introduction

The recovery of crystal defects produced after irradiation or deformation has been extensively studied because of its interest in both fundamental and applied research. The positron annihilation spectroscopy is becoming nowadays a rather routine technique in the study of lattice defects and its utility in the detection of vacancies, voids and other vacancy-type defects has been widely proved (PETERSEN).

Many important defect properties in crystalline materials have been obtained by using positron measurements; however, little attention has been paid to the investigation of defect kinetics in the rare-earth metals.

In this work we report the preliminary results of a positron study of defects in deformed gadolinium.

2. Experimental

Gd specimens (3 N purity) were supplied by Goodfellow Metals in the form of sheets 0.5 mm thick. The plastic deformation was performed at room temperature by cold rolling to a thickness reduction of 22%.

Positron annihilation measurements were carried out at room temperature between 290 and 700 K at intervals of about 50 K after isochronal annealings of 30 min in a nitrogen atmosphere. Lifetime spectra were recorded by using a conventional fast lifetime spectrometer with a time resolution of about 315 ps (FWHM); Doppler broadening measurements were made with a stabilized Ge(Li) detector having an energy resolution of 1.5 keV at 511 keV.

As a reference sample a Gd specimen annealed at 1170 K during 3 hours was used.

After subtracting two source components (Ni foil and salt contributions), one-component analysis gave satisfactory fits. The S-parameter was calculated as the relative area of a fixed region taken around the annihilation peak.

---

[1]) Contributed paper presented during the European Meeting on Positron Studies of Defects, March 23–27, 1987, Wernigerode, G.D.R.

3. Results and discussion

The lifetime value after annealing at 1170 K is $\tau = 230$ ps. Lifetime measurements have been performed earlier in Gd (Rodda, Stewart; Hsu, Liu; Gómez, Serna); however, no clear conclusions about the positron lifetime value in the defect free Gd can be drawn from these measurements. Theoretical calculations (Welch, Lynn) report a value of $\tau = 225$ ps for bulk Gd, very close to our value for the annealed sample.

Milstein and Baldwin recommend a heat treatment to a temperature of 1000 to 1200 K during a period of one hour for recrystallizing a 27% deformed Gd sample. According to these authors and bearing in mind the calculated value for the lifetime in bulk Gd, we attribute the measured lifetime $\tau = 230$ ps in annealed Gd to the value characteristic of bulk material.

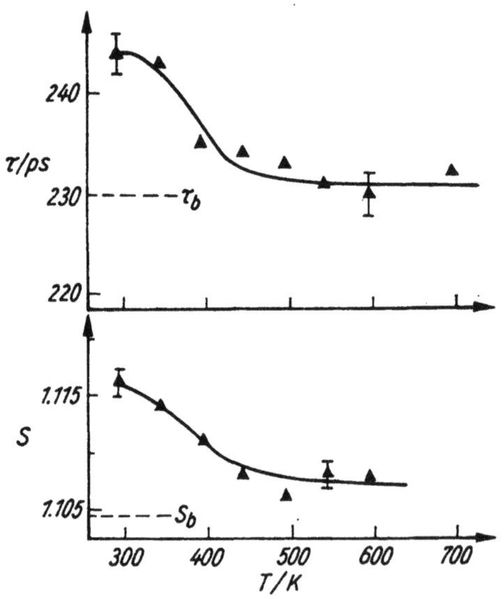

Fig. 1. The evolution of the lifetime τ and the Doppler parameter S as a function of the annealing temperature in a deformed Gd sample. The values marked τ_b and S_b correspond to the positron parameters in a well annealed Gd specimen

The Figure 1 shows the positron lifetime and the S Doppler parameter for the deformed Gd sample as a function of the annealing temperature. The values τ_b and S_b correspond to the measurements in the annealed sample. As it can be seen from the figure the lifetime value after deformation is $\tau \cong 245$ ps. The lifetime shows a decrease up to about 550 K where it reaches the level of 230 ps for the well annealed sample. The lineshape parameter S decreases as well and at 600 K it has almost reached its value before deformation. This indicates that the defects created by cold-rolling recover at a temperature of about 370 K.

Cold work is known to produce dislocations and point defects. The recovery of point defects in electron and neutron irradiated 3 N purity Gd has been studied by Cope et al. and they found that most of the defects produced during irradiation anneal out below room temperature. Since vacancies are created during irradiation, the latter results indicate that also vacancies should be mobile below room temperature, leading to a process of clustering in the deformed metal at room temperature.

As the value of the positron average lifetime after deformation, $\tau \cong 245$ ps, is extremely low to be attributed to three-dimensional vacancy clusters, we assign it to the lifetime of positrons trapped at dislocations. With this interpretation the observed recovery step is attributed to the annealing of dislocations produced by the deformation in agreement with MILSTEIN and BALDWIN, who estimate the start of recrystallization in Gd at 470—570 K.

The most remarkable fact of the present results is the low value of the lifetime for the positrons trapped at dislocations, which is only 6% higher than the bulk lifetime. The positron lifetime associated to dislocations has been found in many metals to be very close to the one related to vacancies, which is usually between 15 and 25% higher than the bulk lifetime.

The decrease observed in the positron average lifetime and S-Doppler parameter for a deformed CuGe alloy at temperatures below the annealing temperature for dislocations has been attributed to impurities atmospheres in the surroundings of the dislocations (HAUTOJÄRVI et al.); more recently a decrease in the positron lifetime associated to dislocations with the purity has been observed in deformed Al (HIDALGO et al.). These results suggest that the low lifetime for positrons trapped and annihilated at dislocations in Gd can be interpreted as a reduction of the open volume associated to dislocations due to impurities anchored to them.

We therefore conclude that the lifetime associated to dislocations in Gd is probably higher than the one observed in the 3 N purity samples, where impurities reduce the dislocation open volume. For the studied samples the anneal of dislocations takes place at about 370 K.

References

COPE, R., MINIER, C., BILGER, H., MOSER, P.: Phys. Lett. **28A**, 510 (1969)
GÔMEZ, M., SERNA, J.: phys. stat. sol. (a) **83**, K131 (1984)
HAUTOJÄRVI, P., VEHANEN, A., MIKHALENKOV, V. S.: Solid State Commun. **19**, 309 (1976)
HIDALGO, C., GONZÁLEZ, G., LINDEROTH, S. (to be published)
HSU, F. H., LIU, H. S.: in Positron Annihilation. Eds P. G. COLEMAN, S. C. SHARMA and L. M. DIANA, Amsterdam 1982, p. 719
MILSTEIN, F., BALDWIN, J. A.: J. Inst. Metals **100**, 337 (1972)
PETERSEN, K.: in: Positrons in Solid State Physics. Eds W. BRANDT and A. DUPASQUIER. Amsterdam 1983, p. 298
RODDA, J. L., STEWART, M. G.: Phys. Rev. **131**, 255 (1963)
WELCH, D. O., LYNN, K. G.: phys. stat. sol. (b) **77**, 277 (1976)

(Received June 16, 1987)

Authors' address:

J. DEL RÍO, N. DE DIEGO, C. HIDALGO
Departamento de Fisica de la Materia Condensada
Faculatd de la Ciencias Fiśicas, Universidad Complutense
E-28040 Madrid, Spain

Book Review

J. LEKNER

Theory of Reflection of Electromagnetic and Particle Waves

Developments in Electromagnetic Theory and Application; 3

Martinus Nijhoff Publishers, Dordrecht/Boston/Lancaster, 1987, 279 S. Dfl. 175.00, US $ 78.50, ISBN 90-247-3418-5

Neuere Entwicklungen in der Ellipsometrie gaben den Anstoß zu diesem Buch, das sich in erster Linie an Ingenieure wendet, die in ihrer Arbeit in der Optik, der Mikrowellen- und Radiotechnik mit Reflexion und Transmission zu tun haben. Demzufolge basiert die Darstellung im wesentlichen auf der klassischen Elektrodynamik. Jedoch werden gleich im ersten Kapitel der Zusammenhang zwischen elektromagnetischen s-Wellen und der Schrödingergleichung sowie die Analogie zwischen elektromagnetischen p-Wellen und akustischen Wellen dargelegt. Zusätzlich werden im Anhang die Hauptresultate für s-Wellen in die „quantenmechanische Sprache übersetzt". Damit wird ein weiterer Kreis von Interessenten angesprochen, Physikern, Chemikern und Ingenieuren, die Reflexion und Transmission von Teilchen an Potentialbarrieren zu behandeln haben. Die vorgestellten Techniken sind darüber hinaus von Interesse für Akustik, Ozeanographie und Seismologie. Bei Zugrundelegung der linearen klassischen Elektrodynamik in unmagnetischen Substanzen und der entsprechenden Teilchenanalogie sind nichtlineare und Quantenoptik nicht Gegenstand des Buches. Der Vielfachstreumechanismus als mikroskopische Grundlage wird in der Einleitung kurz angesprochen. Im übrigen werden die Substanzeigenschaften durchgängig durch eine frequenz- und ortsabhängige dielektrische Funktion bzw. ein effektives Potential beschrieben; beide werden als gegeben vorausgesetzt. Im Rahmen dieser Einschränkungen wäre allerdings die Einbeziehung einer nichtlokalenBeschreibung sehr wünschenswert gewesen. Im wesentlichen werden Schichtprobleme mit ebenen Grenzflächen, auch mit Rauhigkeiten behandelt. Bei der vorgenommenen Einschränkung des Gegenstandes gelingt dem Autor eine anwendungsbezogene, gut lesbare Darstellung, die altbekanntes verbindet mit anspruchsvollen, methodisch gut dargestellten Techniken. Variationsverfahren, Differentialgleichungen für Reflexionsamplituden, Anisotropie werden behandelt, desgleichen inverse Probleme und Impulse. Besonders gelungen und wichtig sind die Behandlung der Dämpfung und das Auftreten der geschwächten Totalreflexion (ATR) sowie Matrixmethoden für Vielschichtprobleme. Insgesamt ist dies ein gutes und wichtiges Buch, das sicher viele Interessenten finden wird.

G. PAASCH

Cryst. Res. Technol.	22	1987	12	1537–1540

Zou Liu-juan, Zhang Cai-guo, Wang Gui-lan, Wang Ming-Zhong

Huazhong University of Science and Technology, Department of Physics, and
Wuhan University, Department of Physics, and
Wuhan Institute of Iron and Steel Technology, Department of Metals, Wuhan, P.R. China

A Research into Positron Annihilation of the B_2F Steel Rolled out in Step-S with Constant Stretch[1])

The rolling out of step-s with constant stretch (PV-E) is a new technology developed for rolling steel in recent years. We studied the B_2F steel rolled by the new technology with the aid of the positron annihilation technique. The relationships of the average lifetime of positron annihilation with the rolling pressures, the ratio of rolling speed and the mechanical properties are obtained. The value of the lifetime shows that there are many dislocations and vacancy clusters in B_2F steel after being rolled, and which are supported by observations with a transmission electronmicroscope.

1. Introduction

The positron annihilation technique has been greatly developed in recent years (Hautojärvi). Its extensive applications are involved in solid physics, metal physics, and material science et al. The information of defects and microscopic structures in the condensed matters can be obtained with the help of positron annihilation technique because the positrons are extremely sensitive to the defects (for example, vacancies, dislocations and microscopic voids, et al.).

The research on the mechanical behaviour of products rolled the out in step-s with constant stretch (PV-E) is an important part for popularizing and applying the new technique, which plays an important role in the improvement of product quality and productive capacity.

This paper tries to study the relations between the mechanical behaviour and microscopic defects of the B_2F band steel rolled by PV-E with the aid of positron annihilation technique.

2. Experiments and results

We chose five groups of B_2F band steel as samples. The thickness, 0.5 mm, of these original sample changed into 0.375 mm after cold reduction, the ratio (λ) of rolling out was 1.32 and implemented the roll with different out of step ratios. The roll-diameters were $64.5 \sim 90.0$ mm, the lengths of the roll body were 150 mm, the material of the roll body was 40 Cr. We also used the nonlubricating roll and took the 7# sample rolled in steps for a comparison.

[1]) Contributed paper presented during the European Meeting on Positron Studies of Defects, March 23–27, 1987, Wernigerode, G.D.R.

Table 1
The components and contents in B_2F steel

components	c	Si	Mn	p	s
contents %	0.09 ~ 0.15	0.12 ~ 0.30	0.25 ~ 0.55	$\leqq$ 0.045	$\leqq$ 0.050

Table 2
The rolling pressure p; ratio, a, of speed and resist strain intensity σ_b and plasticity δ% (WANG GUI LAN)

Sample No	7	9	1	2	8	5
rolling pressure p/kg	5019	2329	3113	3273	3621	3945
ratio, a, of speed	1.32		1.0	1.24	1.41	1.32
resist strain intensity σ_b	59		48	50	51	55
plasticity δ%	5		6.0	11.1	9.0	10.0

Table 3
Positron lifes τ_2 and intensities I_2 of defect capture states for several samples

Sample No	7	9	1	2	8	5
τ_2	254	246	250	254	264	251
I_2	51.6	53.5	53.4	53.2	49.4	53.0

The components and contents in B_2F steel are listed in Table 1. The crystal structure is body centered cubic, the solid solubility gap in α-Fe is carbon.

The rolling pressure P; and the ratio a of rolling speed; the resist strain intensity σ_b and plasticity δ% measured after being rolled are indicated in Table 2.

A positron lifetime spectrum instrument with ORTEC FAST-FAST coincidence was applied in our experiments. The time-resolved ratio, 280 ps, of the system measured under the condition of Na^{22} energy window with radiological source Co^{60}. The positron source (Na^{22}) intensity was about 10 μc. The measurements were done at room temperature, 24 $\pm$ 1 °C, the total integral countings each sample were higher up to 10^6. A simplified POSITRONFIT EXTENT program was used to fit the three lifetimes, τ_1, τ_2, τ_3 and calculated the corresponding (to τ_1, τ_2, and τ_3) intensities I_1, I_2, and I_3, the correction to the source was not done. The positron lifetimes τ_2 and intensities I_2 of the defect capture states are shown in Table 3.

3. Analysis and discussion

1. We know that the characteristic lifetime of the positron free annihilation in Fe is 110 ps and in α-Fe the positron life, τ_2, is 170 ps and $\tau_2 = 240 \sim 300$ ps (HAUTOJÄRVI et al.) for a vacancy capturing and for a cluster of vacancies capturing, respectively. However, we obtained the $\tau_2 = 246 \sim 264$ ps (see Tab. 3) from the B_2F samples, this shows that a lot of defects as dislocations and vacancy clusters et al. exist in that samples whether the samples are rolled in step of PV-E. And the above conclusion is supported by observations with the aid of transmission electronmicroscope (Fig. 1). So, the positron annihilation technique can provide the microscopic information for PV-E rolling.

2. In order to further investigate the relations between the PV-E rolling and positron lifetimes, we plot the curves (Fig. 2) for the variations of τ_2, I_2 with the rolling pressure P.

It can obviously be seen that the variation of τ_2 with p is opposite to the I_2 with p; this shows that the intensity is decreased as the defect size increasing; and τ_2 is increased with p increasing for $p < 3600$ kg, this shows that the microscopic defects

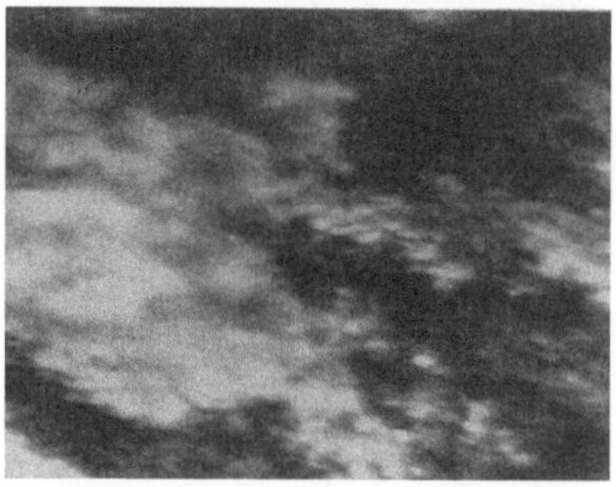

Fig. 1. A photograph of transmission electromicroscope

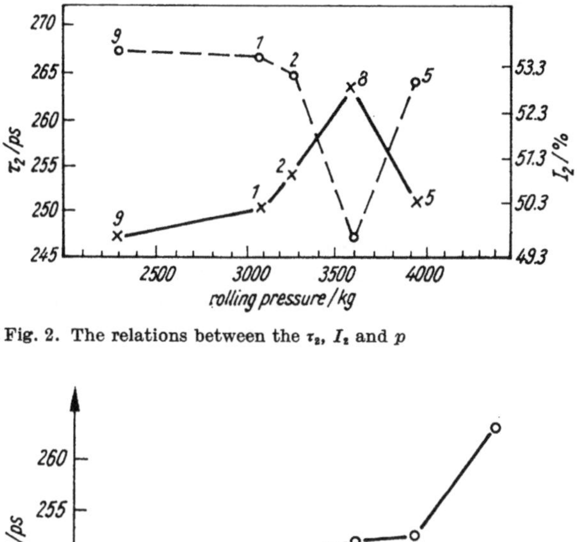

Fig. 2. The relations between the τ_2, I_2 and p

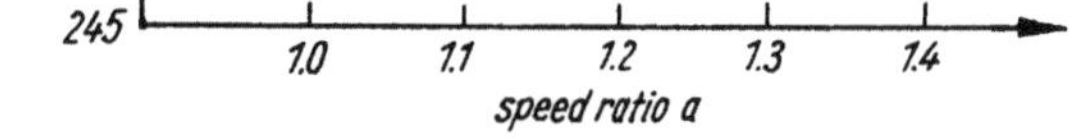

Fig. 3. A relation between the τ_2 and speed ratio a

in the product are related to p in the process of rolling; and τ_2 is decreasing with p increasing for $p > 3600$ kg; this shows that the defects in this product are not monotonically related to p.

3. A curve describing the relation between the τ_2 and speed ratio a is given in Figure 3.

In Figure 3 one finds that τ_2 is increased with the speed ratio a increasing, this represents the corresponding relationship between the speed ratio of rolling and microscopic defects.

4. A relation between τ_2 and plasticity $\delta\%$ is shown in Figure 4. We find that the curve may be divided into two regions. In the I-region, the $\delta\%$ is increased with τ_2 increasing and then tends to a constant value; this is because the deformative storage energy and remainder stress exist in B_2F steel and are against an effect of defects on the plasticity although there are many dislocations and vacancy clusters in B_2F steel rolled with PV-E. In the II-region, the plasticity $\delta\%$ is slightly decreased with τ_2 increasing; the reasons are that the crystal rotation exists in the deformation after PV-E rolling, and some defects extend to form the micro-slight cracks, so, the plasticity of the material is destroyed.

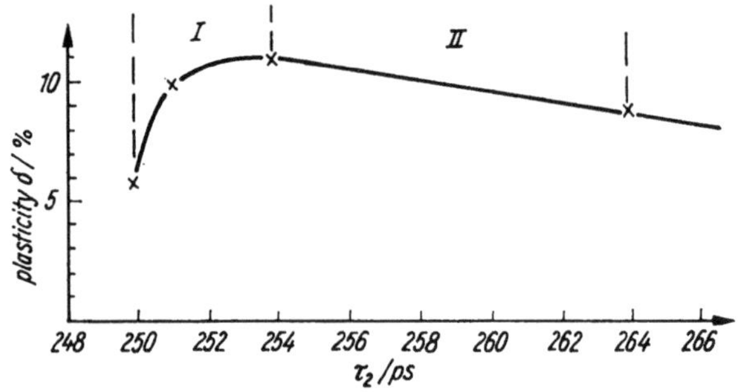

Fig. 4. A relation between τ_2 and plasticity $\delta\%$

4. Conclusions

This research into PV-E rolling the products, with the help of positron annihilation technique, shows that there are many dislocations, vacancies et al. in the B_2F steel, and these corresponding relations between the defects and rolling pressure, speed ratio are found. After PV-E rolling, the mechanical properties, for example the plasticity and relation-ship between the plasticity and microscopic structure in B_2F steel, are found to have certain laws. So the positron annihilation technique can also be used as a tool to investigate the mechanical behaviour in rolling industry.

References

HAUTOJÄRVI, P. (ed.): Positrons in Solids, Berlin 1979

HAUTOJÄRVI, P., JUDIN, T., VEHANEN, A., YLI-KAUPPILA, T., JOHANSSON, J., VERDONE, J., MOSER, P.: Solid State Commun. **29**, 855 (1979)

WANG GUI LAN: Master Thesis, A study on the mechanical properties of a product rolled out in step-s with constant stretch, to be published.

(Received June 16, 1987)

Authors' address:

c/o ZOU LIU-JUAN
Huazhong University of Science and Technology
Department of Physics
Hubei
Wuhan
P.R. China

Cryst. Res. Technol.	22	1987	12	1541–1545

QIAN YANG, MA RUZHANG, HE YONGSHU, HUANG MAORONG,
LIU SHOUDAN

Dept. of Metal Physics, Beijing University of Iron and
Steel Technology, Beijing, China, and
Institute of High Energy Physics, Academia Sinica, Beijing, China

Positron Annihilation Study of Defects in Martensite Transformation of Fe-Mn Alloy[1])

It has been observed that a small number of repetitions of the transformation by thermal cycling enhanced the formation of ε-Martensite whereas a large number of repetitions of transformation decreased the amount of ε-Martensite formed. PAT analysis and electron microscope show that a few repetitions of the transformation produced less defects, including vacancies and dislocations, whereas a large number of repetitions of transformation produced progressively more defects. The PAT analysis also shows that these accumulated defetcs are mainly dislocations, which form a network of dislocations or a Lomer-Cottrell dislocation net. A small number of dislocations enhanced the nucleation of ε-Martensite whereas a large number of dislocations prevented the nucleation. So the dislocations-produced by cycling transformation thermally — has a great effect on the ε-formation.

Es wurde beobachtet, daß die Umwandlung bei einer kleinen Anzahl von Wiederholungen des thermischen Zyklus die ε-Umwandlung zwar gefördert wird, daß jedoch bei mehrmaliger Wiederholung eine Verminderung des ε-Martensit bewirkt wird. Die Ergebnisse der PAT-Analyse und der elektronischen Untersuchung zeigten, daß die Umwandlung bei wenigen Wiederholungen die schädlichen Struktur-Fehler, einschließlich Leerstellen und Versetzungen, reduziert, während bei mehrfacher Wiederholung allmählich mehrere Fehler auftauchen.

Die PAT-Analyse zeigte ferner, daß es sich bei solchen sich häufenden Fehlern hauptsächlich um Versetzungen handelte, wobei ein Versetzungs-Netzwerk oder eine Lomer-Cottrell-Versetzung entstand. Bei wenigen Versetzungen wurde die Keimbildung des Martensit begünstigt, wäjrend bei mehreren Versetzungen die Keimbildung verhindert wird. Also übt die durch thermischen Zyklus der Umwandlung verursachte Versetzung einen großen Einfluß auf die ε-Bildung aus.

Introduction

It has been well established that Martensite can be formed on cooling in Fe–Mn alloys of appropriate composition and various kinds of defects such as twins, stacking faults and vacancies can be produced, due to the dislocation interactions.

At present, several hypotheses about ε-nucleation have been suggested, among which two ones seem more acceptable: kinematical screw mechanism (ZUYAO) and dynamical mechanism (BOLLMANN). Both are in agreement that the embryos of ε (H.C.P.) are stacking facults in B.C.C. (austenite).

In investigations on Fe-20% Mn alloy MA RUZHANG et al. found that Ms = 110 °C, As = 180 °C, Af = 220 °C, and observed that cycling transformation thermally has

[1]) Contributed paper presented during the European Meeting on Positron Studies of Defects, March 23–27, 1987, Wernigerode, G.D.R.

a great effect on the ε-formation (Fig. 1). The author assumed that it is because the repeats of transformation produced defects progressively in austenite and that the few, solitary dislocations are the favourable site of ε-nucleation whereas a large amount of dislocations arisen from a large number of cycling transformation can seriously interfere the growing of ε-embryos, and in turn decrease the amount of ε-phase, which is due to that a large amount of defects and dislocations can form complicated dislocations nets.

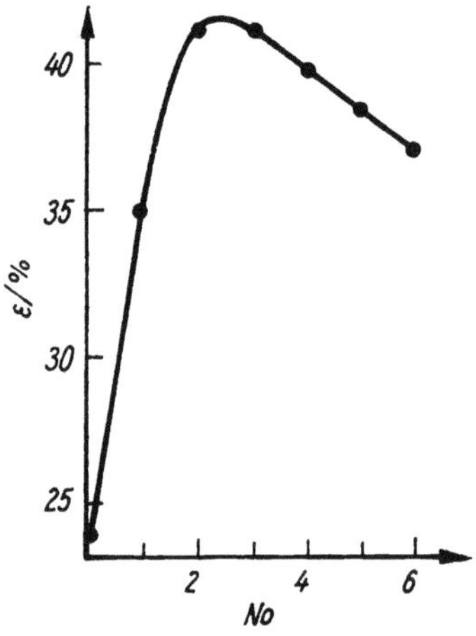

Fig. 1. Number of transformations vs phase quantity

Positron Annihilation Technology is sensitive tool for detecting lattice defects (HAUTOJÄRVI; SEEGER; ZENJEI et al.). It has been used to study the defects produced by martensite transformation in Fe–Ni alloys (YONGSHU et al.), revealing some valuable results.

In this paper, we apply PAT to study the defects and its relationship with the number of cycling transformation in Fe–Mn alloy, aiming to confirm the suggestions proposed by MA RUZHANG et al.

Experimental methods

The composition of alloys is the same as those used by MA RUZHANG et al.: 0.03% C, 0.07% S, 0.07% Si, 0.01% P, 20.81% Mn, and balanced with Fe. The cold-rolled stripes were cut into samples of square shape of 10 × 10 mm and 1 mm thick. The samples were heated to 900 °C for 1 hr. After furnace-cooling, they were annealed to 300 °C for 15 min and then air-cooled. Cycle this process for 10 times to get specimens of repeated transformation for measurement. Finally, the specimens were electron-polished down to 0.5 mm thick. At room temperature, they were measured both Doppler-broadening and positron lifetime for each sample. The specimens were annealed in different temperatures of 250, 350, 400, 450, 500, 600, 700, and 800 °C for 30 min, respectively. Also the positron parameters were measured for each specimen.

In positron parameter measurement, the ^{22}Na was used as a positron source. The lifetimes were measured by positron lifetime spectrometer and the *s* was measured by line broadening. In both the Doppler-broadening and positron lifetime measurement, the total counts of each spectrum were more than 7 × 100000.

The two peculiar specimens, annealed at 250 °C and 800 °C respectively, were observed by transmission electron microscope (H-800 TEM with an accelerating voltage at 200 KV). Thin foils were prepared by electro-polishing in 5% perchloric acid and 95% alcohol.

Experimental results

The relationship between the numbers of cyclic transformation and positron parameters in shown in Figures 2, 3. It can be found there that both the $\bar{\tau}$ and s have a varying trend with the number of cycling transformation, i.e., increasing with the number of cycling transformation. Considering the relationship between the parameters and the defects, this means that the density of defects increase with the number of cycling transformation. This phenomenon can be explained in the following way. Annealing at 900 °C which is much greater than the Ms (110 °C) of Fe-20% Mn alloy will surely bring about $\nu \underset{\text{heating}}{\overset{\text{cooling}}{\leftrightharpoons}} \varepsilon$ transformation vacancies and dislocations. When

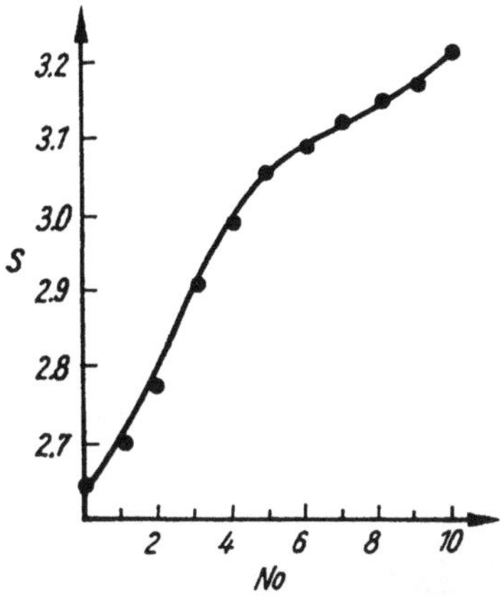

Fig. 2. Number of transformations vs S parameter

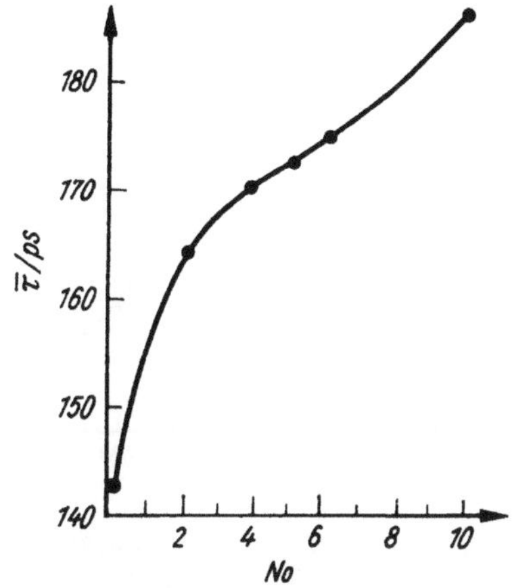

Fig. 3. Number of transformations vs positron lifetime

annealing to 300 °C, the defects caused by previous transformation still exist and at the same time the transformation happens again. Recycling the process will accumulate defects, increasing the $\bar{\tau}$ and s. We also are convinced that the defects are mainly composed of dislocations, because the energy provided by annealing to 300 °C is well enough to delete the vacancy defects, but less enough to delete the dislocation. Further experiments are conducted to confirm the proposal.

When comparing the curves in Figures 1—3, it is also found that the increasing rate of the density of dislocation falls behind that of the ε-phase quantity. It is known that in some conditions, some defects with certain configuration, such as crystal interfaces, and subcrystal interfaces which are composed of partial dislocations or dislocations, are favourable to nucleate martensite. But the defects of intensified parent phase, such as dislocation net, dislocation jog will prevent the dislocation motion and the growing of martensite embryos (ZUYAO). So the formation of a small amount of dislocation defects produced by a small number of cycling transformations is responsible for the rapidly increased ε-phase because it can decrease the nucleation energy. According to the ε-nucleation theory, the expanding edge of dislocation is exactly the site of formation of ε-embryo (FUJITA et al.; MIRZAYEV et al.; VOTAVA).

When the number of cycling transformations is between 3—10 times, the density of dislocation still increases whereas the amount of ε-decrease. Responsible for this phenomenon are the defects accumulated progressively by the repeats of cyclic transformation, which formed dislocation roll or even the complicated dislocation net due the interaction of dislocations, which further hindered the motion and expanding of stacking faults. Consequently, the amount of ε was cut down obviously. This result is consitent with the suggestion proposed by MA RUZHANG et al. that the effect of repeats of transformation on ε-phase is due to the piling up of defects. In addition, the progressively accumulated dislocation also increase the non-chemical resistance, against which a larger difference of free energy is needed. This made it difficult to form transformation, and consequently the amount of the defects decrease. But because after each transformation cycle, there is always a positive addition of defects, the total amount of defects still increases, although in a more slower rate than that of those few number of cycling treated samples, as noted in the Figures 1–3.

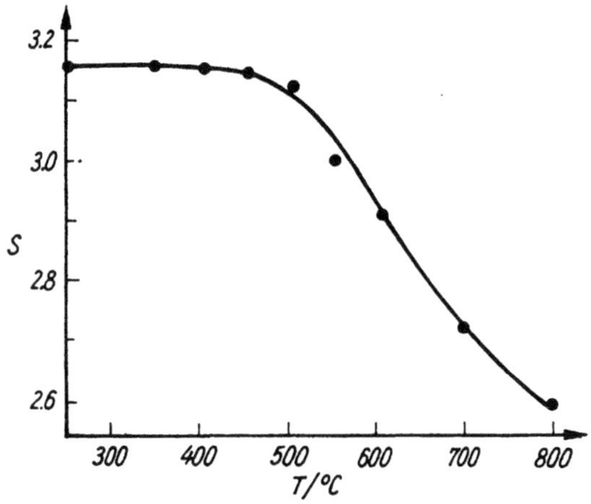

Fig. 4. Heating temperature vs S parameter

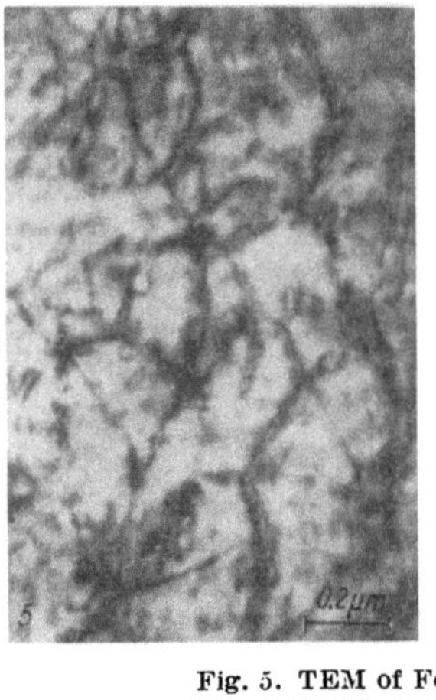

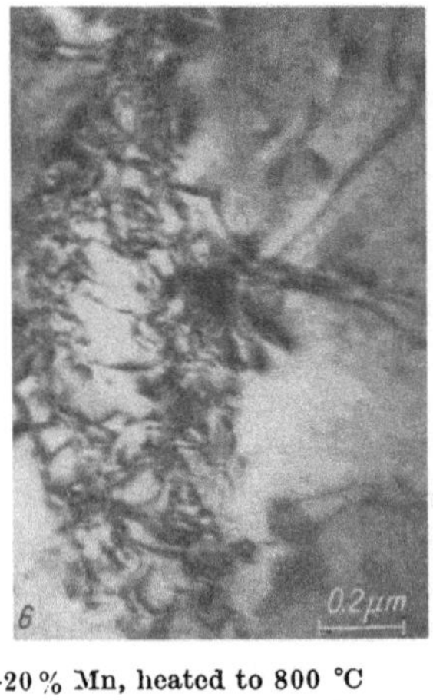

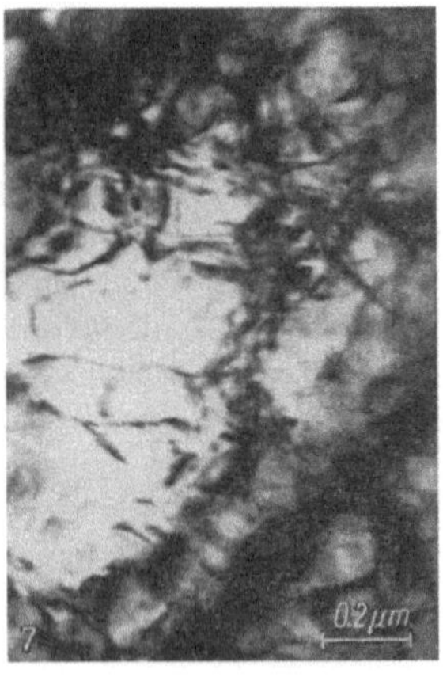

Fig. 5. TEM of Fe-20% Mn, heated to 800 °C
Fig. 6. TEM of Fe-20% Mn, heated to 250 °C
Fig. 7. TEM of Fe-20% Mn, heated to 250 °C

Further experiments are conducted to verify the above hypothesis. The specimens, after 10 cyclings of phase transformation, were heated at different temperatures. The positron parameters were measured at room temperatures for each sample (Fig. 4). It shows below 450 °C, the curve is approximately straight, indicating that the defects don't change in this temperature range. Above 450 °C, the curve turns down, meaning that at 450 °C, dislocations begin to vanish. Between 500–600 °C, dislocations disappear most fastly, for when heating to 450 °C, it can provide enough activation energy to dissolve the dislocations accumulated by 10 cyclings of phase transformation. Besides, the observation of specimens annealed at 250 and 800 °C by TEM shows that on the 250 °C annealed specimen there are concentrated dislocation network whereas on the 800 °C annealed specimen the dislocations are much less than those of the above one. This again prove our hypothesis that the defects are mainly dislocations accumulated (Figs. 5, 6, 7).

Conclusion

The repeats of the transformation by thermal cycling enhance or impede the formation of ε-martensite. A small number of repeats of phase transformation bring about few defects which enhanced the nucleation of ε-phase, while a large number of repeats of transformation produce progressively more dislocations which prevent the nucleation. So the dislocations produced by thermal cycling has great effect on the ε-formation.

The authors wish to thank Miss Yang Jiuhua for her good cooperation. This work was done under the auspices of the Lab of the Dept. of Metal Physics in Beijing University of Iron and Steel Technology & Inst. of High Energy Physics, Academia Sinica.

References

Bollmann, W.: Acta Metallurg. **9**, 972 (1961)
Fujita, H. et al.: Acta Met. Trans. **1**, 1577 (1970)
Hautojärvi, P. et al. (editors): Positrons in Solids, Berlin 1979
Mirsayer, D. A. et al.: fiz. Metal Metalloved **37**, No. 5, (1974) p. 912
Ruzhang, M.: Transcations of Metal Heat Treatment, Vol. 3 (1982), No. 2, p. 27
Seeger, A.: J. Physics, F. **3**, 248 (1973)
Votava, E.: Acta Metallurg. **8**, 81 (1960)
Yongshu, H. et al.: Positron Annihilation, edited by P. C. Jain, R. M. Singru and K. P. Gopinathan, Singapore, 1985, p. 611
Zenjei, W. et al.: Acta Metallurg. Sin. (China) **17**, No. 6 (1981) p. 613
Zuyao, X.: Martensite and Martensite Transformation, 1980

(Received June 16, 1987)

Authors' address:

c/o Qiang Yang
Department of Metal Physics
Beijing University of Iron and Steel Technology
Beijing
P.R. China

Book Review

A. Niku-Lari

Advances in Surface Treatments

Technology — Applications — Effects — Volume 3

Pergamon Press, Oxford-New York-Beijing-Frankfurt-Sao Paulo-Sydney-Tokyo-Toronto, 1986, 260 Seiten, 240 Abbildungen, 25 Tabellen, Preis: US $ 61.00, ISBN 0-08-033 464-4

Durch geeignete Oberflächenbehandlungsverfahren gelingt es, den Widerstand von Bauelementen gegen Ermüdung, Verschleiß und Korrosion zu erhöhen. Dadurch sind erhebliche ökonomische Vorteile erreichbar. Es ist das Anliegen des Herausgebers, potentiellen Nutzern von Oberflächenbehandlungsverfahren aktuelle wissenschaftliche und technische Informationen zukommen zu lassen. In Fortsetzung der bislang erschienenen beiden Bände zur gleichen Thematik ordnet der Herausgeber im vorliegenden Band 3 insgesamt 21 Aufsätze folgenden Themengruppen zu:

- Einfluß der Oberflächenbehandlung auf Ermüdung, Rißentstehung und Rißausbreitung;
- Eigenspannungen;
- Oberflächenverfestigung und Oberflächenbeschichtung;
- Technologie, Kontrolle und industrielle Anwendung der Oberflächenbehandlungsverfahren.

Eingangs findet man eine übersichtliche Auflistung des Herausgebers über die Hauptgebiete aller in Betracht kommenden Oberflächenbehandlungsverfahren und -techniken. Danach folgt ein Übersichtsbeitrag zu den bei Wärmebehandlungen ggf. auftretenden Deformationen mit den sich herausbildenden Eigenspannungen. Die folgenden 6 Beiträge befassen sich unter teilweiser Einbeziehung bruchmechanischer Betrachtungsweisen vorrangig mit dem Einfluß des Kugelstrahlens auf den Rißbeginn und die Rißausbreitung während schwingender Beanspruchung. Zur zweiten Gruppe gehören 4 Aufsätze, die über Ermittlung und Relaxation von Eigenspannungen Aussagen treffen. In der dritten Themengruppe findet man bemerkenswerte Beiträge zur Oberflächenhärtung durch Ionenimplantation von Bor und Stickstoff sowie über Auswirkungen des Kugelstrahlens auf die Kontaktermüdung. Zur letzten Themengruppe zählen 5 Aufsätze, in denen Erfahrungen über die Auswirkungen des Kugelstrahlens auf das Ermüdungsverhalten von Schraubenfedern und Flugzeugbauteilen mitgeteilt werden.

Die Beiträge vermitteln einen Eindruck vom gegenwärtigen Stand der Erkenntnis bestimmter Oberflächenbehandlungsverfahren.

Bei der Herausgabe weiterer Aufsatzsammlungen zur gleichen Thematik sollten mehr Übersichtsvorträge aufgenommen werden.

G. Schott

Cryst. Res. Technol.	22	1987	12	1547–1549

LIU SHAN-YI, ZHU GUAN-QU, WANG SHAO-JIE

Wuhan Metallurgical Institute, Wuhan, China, and
Department of Physics, Wuhan University, Wuhan, China

Studies on the Relation between the Magnetic Properties and Positron Lifetime of ALNICO 8 Permanent Magnet Alloys

Alloys ALNICO 8 have been wildly used in industries for their excellent magnetic properties and reliable temperature stability.

The present paper has studies the relation between the maximum magnetic energy product (MMEP) of columnar crystallizated AlNiCo 8 permanent magnet alloys cooled continuously in magnetic fields and positron lifetime in this material. The experimental results show that the minimum decay constant τ_1 reaches gradually its saturation as MMEP increase, this means the trapped effect obviously happens in the specimens with low MMEP. With the increase of MMEP, the τ_2 component reflecting the crystal imperfection decreased, hence, the crystal perfection improved. Mössbauer spectroscopic results also show that with the increase of MMEP, the more parts of magnetic moments orientate to magnetizate direction. This work gives valuable data to improve the properties of materials.

1. Introduction

Notwithstanding the appearance of new hard magnetic materials, AlNiCo type alloys still constitute an important group of materials for the production of magnets used in various technical applications (SCHÖLER; SOZKA). Alloys AlNiCo have been excellent magnetic properties and reliable temperature stability. Many problems are solved concerning remagnetization theory in the AlNiCo type alloys. Nevertheless the magnetic properties of the AlNiCo type alloys obtained in practice are far from those theoretically predicted (GRACYANOV). The present paper has studies the relation between the MMEP of columnar crystallizated AlNiCo 8 permanent magnet alloys cooled continuously in magnetic fields and positron lifetime in this material. The aim of the present work is to correlate the achieved magnetic properties with microstructural characteristics.

2. Experimental

We have prepared an alloy of AlNiCo 8 with the following composition: 34% Fe, 35% Co, 14.5% Ni, 7% Al, 5% Ti, 3% Cu (in wt.%) and small amounts of S and Nb. Specimens (cylinders of 18 mm dia and 17 mm length) were casted in a usual way. The alloys were melted by means of 60 kW high frequency induction furnace. The used heat treaterval consisted of homogenization at 1250 °C for 18 min, cooled continuously in magnetic fields and following aging at 650 °C for 5 hours, at 580 °C for 10 hours and at 550 °C for 20 hours.

[1]) Contributed paper presented during the European Meeting on Positron Studies of Defects, March 23–27, 1987, Wernigerode, G.D.R.

The positron lifetime spectra were measured at room temperature by fast-slow coincidence system connected to XP2020 photomultipliers tubes with a 7 μCi ^{22}Na positron source and a time resolution (FWHM) for ^{60}Co of 290 ps. The spectra were analyzed by POSITRONFIT EXTENT program to extract three components. The intensity I_3 and lifetime τ_3 of third component were 3−4% and 1.7 ns, respectively. It can be attributed to annihilation in the source and at surfaces.

The specimens of the Mössbauer spectra were slice smaller than 50 microns. The measurements of the Mössbauer spectra were done with the electrodynamic transducer in the constant acceleration regime. The scintillation γ spectrometer with multichannel analyzer served as a registration system. The source (Co 57 in Rh matrix) was at room temperature.

3. Results and discussion

3.1. The results of positron lifetime spectra are listed in Table 1. As shown in Figures 1 and 2, the change of I_2, τ_2 indicated the magnet characters at different the MMEP. Positron lifetime τ_2 reflects crystallographical imperfection, for example, vacancies,

Table 1
Observation of positron lifetime spectra of AlNiCo8 Alloy

(MMEP) (BH) maxMGOe	parameters of positron lifetime spectra						index of curve fitting
	τ_1/ps	τ_2/ps	τ_3/ps	I_1/%	I_2/%	I_3/%	
7.19	115 ± 1	316 ± 10	1715 ± 32	80.7 ± 1	15.6 ± 0.9	3.7 ± 0.1	0.967
8.10	114 ± 1	327 ± 9	1780 ± 39	82.0 ± 0.8	14.9 ± 0.8	3.1 ± 0.1	1.338
9.56	110 ± 1	310 ± 9	1719 ± 33	81.1 ± 0.9	15.5 ± 0.8	3.4 ± 0.1	1.546
10.63	112 ± 1	309 ± 9	1726 ± 34	80.9 ± 0.9	15.7 ± 0.7	3.4 ± 0.1	1.364
11.34	109 ± 1	298 ± 8	1728 ± 31	78.4 ± 1	18.0 ± 1	3.7 ± 0.1	1.300
12.40	110 ± 1	297 ± 8	1721 ± 31	77.1 ± 1	19.3 ± 1	3.7 ± 0.1	1.021

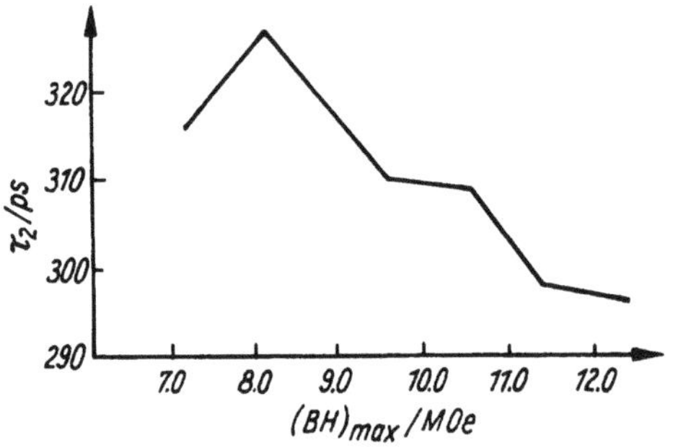

Fig. 1. τ_2 as function of MMEP

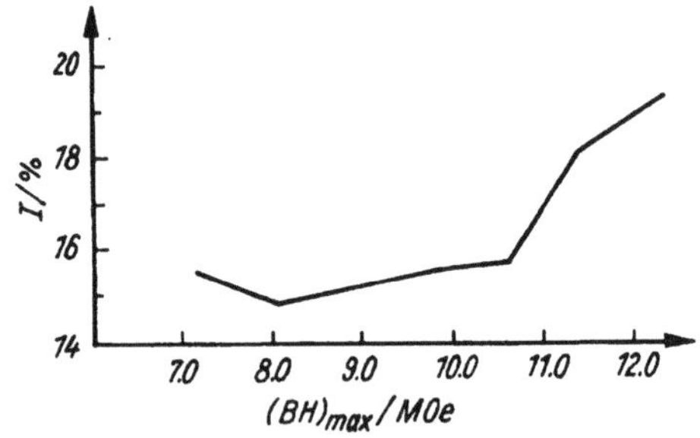

Fig. 2. I_2 as function of MMEP

holes and dislocations. With the increase of MMEP, the τ_2 component reflecting the crystal imperfection decreases, the crystal perfection improved. The high temperature quench process may freeze the defects the distribution of which has not reached equilibrium, and then creates Schottky defects. The annealing may take place on the surface when the crystal is rapidly cooled, and the defects condense, even produce mechanical stress and dislocations. However, after ageing treatment, atoms obtain equilibrium by diffusion and some imperfection disappear. Therefore vacancies decreased and the inhomogeneity from casting process improved.

3.2. The experimental results show that the minimum decay constant τ_1 reaches gradually its saturation as MMEP increases, this means the trapped effect obviously

happens in the specimens with low MMEP. This kind of samples has a high impurity level.
3.3. The Mössbauer spectra of the alloys are presented in Figure 3. The alloys embody six lines of hyperfine magnetic structure. The broadening is attributed to the presence in the alloy of a number of iron atom sites with different nearest surroundings. In its central part an additional line was observed. The presence of this line indicates that

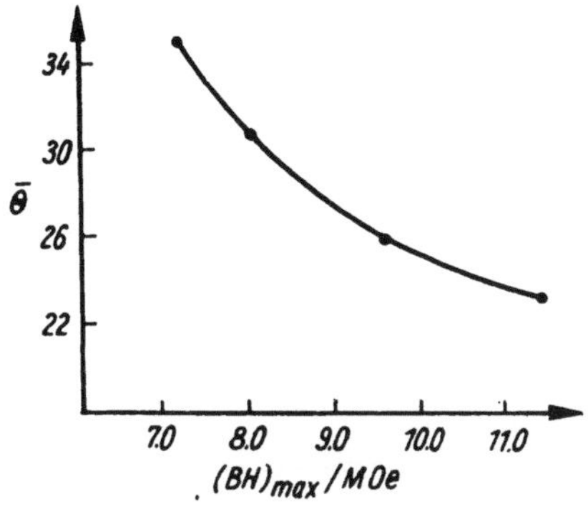

Fig. 3. $\bar{\theta}$ as function of MMEP

some iron atoms are in paramagnetic surroundings. θ is the included angle between r-ray and the direction of magnetization.

$$\overline{\cos^2 \theta} = 1 - \frac{2(I_2 + I_5)}{\sum_{i=1}^{\sigma} I_i}$$

I_i — intensity of i line.

As shown in Figure 3 $\bar{\theta}$ decrement with increase of MMEP, the more parts of magnetic moments orient to magnetization direction.

The optimum magnetic properties depend on crystallographical orientation in casting process and morphology and orientation of α_1 particles are oriented in $\langle 100 \rangle$ matrix directions. The preference of particles formation in the direction of magnetization causes an improvement of magnetic properties.

References

GRACYANOV, Yu. A. et al.: Metallurgye Precizyonnych Splavov Metallurgya, Moscow 1975, p. 283
SCHÖLER, K.: TEW Technische Berichte 3, 41 (1977)
SOZKE, L., SZYMURE, S.: Prace Instytutow Hutniczych 29, 33 (1977)

(Received June 16, 1987)

Authors' addresses:

LIU SHAN-YI, ZHU GUAN-QU
Wuhan Metallurgical Research Institute
12 Long DI, Hanyang
Wuhan, P.R. China

WANG SHAO-JIE
Department of Physics, Wuhan University
Wuhan, P.R. China

Book Review

B. R. Pamplin †, P. Krishna (eds)

Recent Advances in X-Ray Characterization of Materials

Progress in Crystal Growth and Characterization, vol. 14

Pergamon Press, Oxford/New York/Beijing/Frankfurt/Sao Paulo/Sydney/Tokyo/Toronto 1987, 514 p., Bibliography, Subject and Compound Index. DM 300. –

The aim of this volume is to review the latest developments and advances made in the field of X-ray characterization of materials. Indeed, this sphere of materials research has been subjected to such a progress during the last decade that it is justified to call it tremendous (P. Krishna).

The volume covers a wide range of techniques dealing with materials of different properties ranging from highly pure single crystals to totally disordered systems. The book is divided into 12 chapters written by eminent scientists.

The topics treated are: The X-ray anomalous dispersion corrections and their use for the characterization of materials (D. C. Creach); Diffuse X-ray scattering from crystals (H. Jagodzinski); Single crystal diffraction studies of stacking faults in close-packed structures (M. T. Sebastian and P. Krishna); Some applications of X-ray fluorescence analysis (M. Mantler); X-ray diffraction on single crystals at high pressure (H. Ahsbahs); Single crystal X-ray diffraction studies of materials at high temperature (H. Schulz); Characterization of the perfection of large single crystals by means of γ-ray diffractometry (J. R. Schneider and H. A. Graf); X-ray topography of twinned crystals (H. Klapper); Real time and stroboscopic topography (B. K. Tanner and J. K. Cringean); X-ray topographic studies of phase transition in crystals — the α-β transition of quartz (N. Kato and K. Gouhara); The simulation of X-ray topographic images (Y. Epelboin).

The contributions give a detailed picture of the state of research concerning theory and application of X-ray investigations. Nevertheless, some readers could miss one or the other topic. Therefore, a second part is under way completing this issue and underlining once more the significance of the X-ray methods and their broad use in crystallographic research.

This volume can be recommended twice. It provides a good informative survey as a whole and offers the possibility to become familiar with special points of interest to a large degree.

W. Schmitz

Cryst. Res. Technol.	22	1987	12	1551–1555

S. R. Dhanalakshmi, T. Nagarajan, S. Ramasamy, R. N. Viswanath, A. Manivannan, G. V. Subba Rao

School of Physics, University of Madras, Madras, India, and
Materials Science Research Centre, I.I.T., Madras, India

Positron Lifetime Studies in Pure and Chromium Intercalated 2H–$NbSe_2$[1])

The layered 2 H–$NbSe_2$ belongs to a class of very interesting low dimensional conductors which exhibits CDW instability at 32 K. In our earlier paper we have established the two dimensionality of the Fermi surface, in these layered compounds by positron angular correlation technique. The reduction in Fermi level along the basal plane due to CDW transition has been already demonstrated by the Japanese group. But the sensitivity of positron lifetime and Doppler broadening techniques for CDW transition is controversial. The Japanese group report drastic changes in Doppler broadening parameters whereas the American group do not find such a change. Positron lifetime measurements have been performed on pure and Cr intercalated 2 H–$NbSe_2$ with Cr concentration 0.25, 0.33, and 0.5 with a view to assess the origin of the second lifetime component. The results show that there are two lifetime component $\tau_1 \sim 205$ ps and $\tau_2 \sim 440$ ps having intensity $I_2 \sim 29\%$. On Chromium intercalation the intensity I_2 decreases systematically. The trapping rate of positrons shows a systematic decrease with increase of Cr content indicating that the second lifetime arises due to positron annihilation in van der Waal's gap. The defect free annihilation rate λ_f was found to increase on intercalation indicating a charge transfer from the guest intercalates to the host MCh_2 conduction bands.

1. Introduction

The low-dimensional layered compounds are very interesting from the point of view of Fermi surface instabilities resulting CDW transition. This two dimensional character their modification on intercalation resulting from charge transfer has been established by us (Ramasamy, Nagarajan) several years ago. The reduction in Fermi level in basal plane due to onset of CDW transition has been established by the Japanese group. They also report drastic changes in Doppler-broadening parameters due to CDW transition. But Jean and Fluss report that Doppler-broadening as well as lifetime components are not sensitive to low temperature CDW transition. They also conclude that the second lifetime component is due to thermally generated vacancies.

2 H–$NbSe_2$ is a member of the family of layered transition metal dichlacogenides possessing a hexagonal layer structure with a strong intralayer covalent bonding and a weak interlayer van der Waal's bonding. The 3d series of the transition metals from well-defined intercalation compounds with group IV and Vb dichalcogenides. The intercalated atom goes into the octahedral or tetrahedral holes depending on the size, electronic configuration and nature of the host. At specific stoichiometries like ($xs = \frac{1}{4}, \frac{1}{3}, \frac{1}{2}, \frac{2}{3}$ etc.) structural ordering (Fig. 1) of the intercalated atom takes

[1]) Contributed paper presented during the European Meeting on Positron Studies of Defects, March 23–27, 1987, Wernigerode, G.D.R.

place which exhibits super structure of the intercalated product and destroys the 2D order of the lattice. Except for Nickel the maximum stoichiometry of x in Z_xMCh_2 is 0.5. Hughes has constructed the Fermi surface for 2 H–$NbSe_2$ from band structure calculations and has shown that d_z^2 band is half filled and the Fermi level E_F lies near a very sharp peak in the density of states. Later calculations by BULLET shows that d_z^2 band is less than half filled. Band calculations performed by WILSON et al. shows that there is an anisotropy in the Fermi surface. The calculated Fermi surfaces are to the layers.

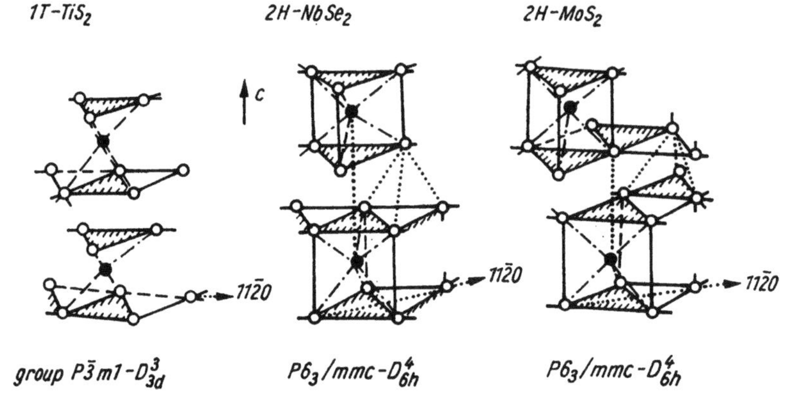

Fig. 1. Structures of MCh_2 (M = Ti, Nb, and Mo) compounds. Open circles: chalcogen, and closed circles: metal

Many group Vb dichalcogenides are known to exhibit intra polytypic phase transitions associated with super lattice formation. These CDW transitions change the neutron and electron diffraction patterns and anomalies in magnetic susceptibility and electrical transport properties of 2 H–$NbSe_2$ which has a CDW transition at 32 K.

Positron lifetime and Doppler broadening studies reported by HATANO et al. (a) on $TaSe_3$, $NbSe_3$, and $TaSe_2$ shows that the metallic compounds have a shorter lifetime than semiconducting compounds and they have found large anisotropies in the two spectra of different crystal direction. They have concluded that the momentum distribution is narrower in the layers than perpendicular to the layers in low dimensional conductors. Angular correlation results of 2 H–$NbSe_2$ reported (HATANO et al. (b)) by the same group at room temperature and 25 K shows that the cut off momemtum along layer direction is smaller than — to the layers. They have concluded that positron annihilation techniques are sensitive to study Fermi surfaces especially when there is an anisotropy in their shape. Their results of the measurements at 25 K show that the anisotropy observed at room temperature reduces below the CDW phase transition. The Doppler-broadening results reported for 1 T–TaS_2 measured between 77 and 292 K in the range of its CDW phase transition temperature 200 K shows that there is an anisotropy in electron momentum distribution. They have investigated the temperature dependence of W parameter for the $C_\perp$ and C 11 directions which shows that there is a steep increase of W parameter in the $C_\perp$ direction 3% comparable to 0.3% in the C 11 direction. Recently positron lifetime measurements for 2 H–$TaSe_2$ reported by JEAN and FLUSS for the temperature range 10—78 K have concluded that the short lived lifetime component $\tau_1 \sim 173$ ps is attributed to annihilation in perfect lattice and long lived lifetime component $\tau_2 \sim 378$ ps to thermally generated vacancies. For the same sample the Doppler broadening measurements reported (JEAN, FLUSS) in the temperature range 20—300 K po not show any abrupt change at CDW transition but indicates anisotropy lattice expansion.

NAITO and TANAKO have reported their results of resistivity measurements made on 2 H–$TaSe_2$ and have concluded that there is an increase in resistivity which is uniform and the Hall coefficient is positive and stays constant. But if there is thermally generated vacancies in 2 H–$TaSe_2$ as reported by JEAN and FLUSS the resistivity cannot increase and the Hall coefficient also cannot remain constant.

In the light of these controversies about layered compounds we have chosen 2 H–$NbSe_2$ for which an anisotropy in Fermi surface at room temperature has been reported by our group (RAMASAMY, NAGARAJAN) as well as HATANO et al. (a). We have intercalated it with Cr_x ($x = 0.25$, 0.33, and 0.5). The aim of the present investigation is to find how positron lifetime parameters change on intercalation, since intercalation facilitate 3D-ordering and destruction of layer type characteristic of these compounds and also to verify whether the second lifetime arises from extrinsic defects. Earlier RAMASAMY and NAGARAJAN have cited destruction of 2D character of 2 H–$NbSe_2$ on intercalation by positron angular correlation measurements and have predicted a change transfer from the Cr atoms to Nb conduction band.

2. Experimental

2H–$NbSe_2$ was prepared in crystalline powder form by direct reaction of high purity (99.99%) elements in an evacuated sealed quartz tube at 900 °C for eight days and cooled to room temperature in the furnace. X-ray powder pattern (CuKα radiation) showed the presence of only 2 H–$NbSe_2$. Cr intercalated compounds (Cr_xNbSe_2, $x = 0.25$, 0.33 and 0.5)

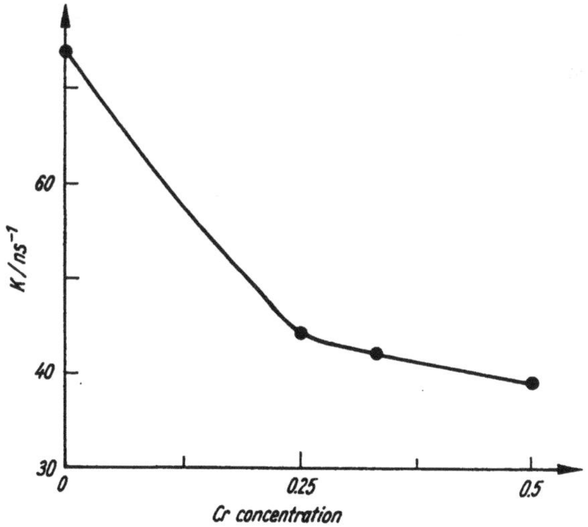

Fig. 2. Capture rate of e^+ vs Cr concentration

were prepared by direct reaction of high purity elements in stoichiometric proportions. The mixtures were heated in an evacuated quartz tube to 900 °C and 1000 °C, respectively, and held at those temperatures for two days. The lattice parameters obtained from X-ray diffraction agreed with those reported in the literature.

Positron lifetime measurements for all the samples have been carried out using a conventional fast slow coincidence system with a time resolution of 240 ps (FWHM). The positron source is sandwitched between two pellets. Each spectrum was measured for

24 to 26 hours during which time about 2 × 10^6 counts were accumulated. The lifetime spectra were analysed after source and back ground corrections as a sum of exponential decay components using POSFIT program. All the samples required two components for good fit.

3. Results and discussion

Positron lifetime results are shown in Table 1. An analysis of this table shows that the short lived component $\tau_1 \sim 215$ ps is comparable to $\tau_1 \sim 233$ ps reported by HATANO et al. (a) for $NbSe_3$. The long lived component in the present measurement is

Table 1
Positron lifetime results of pure and Cr intercalated 2 H–$NbSe_2$

Sample	τ_1 ps	τ_2 ps	I_2 %	λ_f $(ns)^{-1}$	K $(ns)^{-1}$
2 H–$NbSe_2$	207 ± 3	441 ± 7	29	3.64	74.0
$Cr_{0.25}NbSe_2$	216 ± 3	437 ± 12	19	3.88	44.4
$Cr_{0.33}NbSe_2$	215 ± 3	457 ± 13	17	3.91	41.8
$Cr_{0.5}NbSe_2$	214 ± 3	453 ± 13	16	3.97	39.4

$\tau_2 \sim 440$ ps against $\tau_2 \sim 499$ ps reported by them. The magnitude of lifetimes reveals that the compounds under investigation were metallic in nature. In pure 2 H–$NbSe_2$ the intensity I_2 of the second lifetime component is 29%. On intercalation with Cr to a concentration of 0.25, I_2 reduces from 29% to 19%. On further increase of Cr to 0.33 and 0.5, I_2 decreases to 17% and 16%, respectively. This clearly shows that the octahedral holes present in these compounds are progressively filled upon intercalation. This is consistent with the angular correlation results of RAMASAMY and NAGARAJAN wherein they have observed the destruction of 2D character. On Cr intercalation the intensity I_2 of the second lifetime components decreases systematically. This clearly reveals that the octadehral holes present in these samples serve as efficient trapping sites for positrons and these holes are progressively filled by Cr intercalation. The trapping rate of positrons have been computed which show a systematic decrease with increase of Cr content (Fig. 2). The defect free annihilation rate λ_f was found to increase on intercalation indicating a charge transfer from the guest intercalates to the host MCh_2 conduction bands.

4. Conclusion

The capture rate of positrons systematically decreases on intercalation. The defect free annihilation rate λ_f shows a systematic increase with Cr concentration thereby confirming a charge transfer from Cr atoms to the Nb conduction band. Hence it is concluded that the van der Waal's gap present in these layered compounds act as intrinsic positron traps which reduced on intercalation.

Work supported by UGC COSIST INDIA.

References

BULLET, D. W.: J. Phys. C. Solid State Phys. **11**, 4501 (1978)
HATANO, T., SUZUKI, A., OHTAKE, K., SATO, E., YAMAMOTO, R., DOYAMA, M., ENDO, K., IHARA, H., KANEKO, K., GONDA, S.: Proc. of the ICPA-6, 260 (1982) a
HATANO, T., OHTAKE, K., SATO, E., SUZUKI, A., YAMAMOTO, R., DOYAMA, M., ENDO, K., IHARA, H., KANEKO, Kl. GONDA, S.: Proc. of the ICPA-6, 599 (1982) b
JEAN, Y. C., FLUSS, M. J.: J. Phys. C. Solid State Phys. **17**, 2619 (1984)
NAITO, M., TANAKA, S.: J. Phys. Soc. Jpn. **51**, 219 (1982)
RAMASAMY, S., NAGARAJAN, T.: Physica **100B**, 202 (1980)
WILSON, J. A., DISALVO, F. J., MAHAJAN, S.: Adv. Phys. **24**, 117 (1975)

(Received June 16, 1987)

Authors' addresses:

S. R. DHANALAKSHMI, T. NAGARAJAN, S. RAMASAMY, R. N. VISWANATH
University of Madras, Department of Nuclear Physics
A.C. College Campus
Madras 600025, India

A. MANIVANNAN, G. V. SUBBA RAO
Materials Science Research Centre, I.I.T.
Madras 600036, India

Book Review

H. BARTSCH, J. WOLTERSDORF, H.-CHR. GERSTENGARBE, P. WERNER

Elektronenmikroskopische Querschnittsabbildung

Akademie-Verlag Berlin 1987, 84 Seiten, 47 Abbildungen, 112 Quellennachweise, Preis 12.– M
ISBN 3-05-500 233-4

Diese Veröffentlichung reiht sich als Band 15 in die Reihe „Beiträge zur Forschungstechnologie" ein. Die 84 Seiten umfassende Broschüre vermittelt dem Leser an Hand von Beispielen eine Übersicht über die Möglichkeiten und Grenzen sowohl der Probenpräparation als auch der elektronenmikroskopischen Querschnittsabbildungstechnik. Die vorgestellten Ergebnisse resultieren aus Untersuchungen, die am Institut für Festkörperphysik und Elektronenmikroskopie der Akademie der Wissenschaften in Halle vorgenommen wurden. Gegenstand dieser Untersuchungen sind ionenimplantierte Si-Schichten, epitaktische Element- und Verbindungshalbleiterschichten und Halbleiter/Oxid-Grenzflächen. Einleitend wird in allgemeiner Form auf Probleme und Schwierigkeiten bei der theoretischen Beschreibung von Grenzflächen und grenzflächennahen Bereichen eingegangen. Neben einer Beschreibung des Aufbaus und der Funktion von Ionenstrahlabdünnapparaturen werden im weiteren ausführlicher die Verfahren (face to face- und wire shadowing-Technik) zur Präparation von Querschnitten diskutiert. Anhand von Beispielen wie ionenimplantiertes Silizium, Verwachsungen von Si, GaAs und GaN mit Saphir und Spinell wird die Abbildung von Störungen (Zwillinge, Versetzungen, amorphe und polykristalline Bereiche) im Beugungskontrast und deren Anordnung direkt an der Grenzfläche oder/und im grenzflächennahen Bereich vorgestellt und diskutiert, wobei Ergebnisse von in-situ-Heizexperimenten an Querschnittsproben Berücksichtigung finden. Neben der Abbildung der Kristallstörungen u. dergl. in Querschnitten mittels Beugungskontrast werden ebenfalls Ergebnisse von Hochauflösungsuntersuchungen an Grenzflächen der Art Elementhalbleiter/Oxid und Verbindungshalbleiter/Oxid vorgestellt und deren atomarer Aufbau diskutiert. Dem Leser werden Praxiserfahrungen bezüglich der elektronenmikroskopischen Querschnittsabbildung und Probenpräparation vermittelt und damit Anstöße zur gezielten Nutzung dieser aktuellen Technik gegeben. Eine wertvolle Hilfe für den interessierten Leser sind die zahlreichen Literaturhinweise zu diesem Problemkreis.
Im Fall einer weiteren Auflage dieser Veröffentlichung sollte in Hinblick auf eine richtige Darstellung der Begriff „Gasphasenätzung" in der 9. Zeile von unten auf Seite 55 durch „Gasphasenepitaxie" ersetzt werden.

G. WAGNER

Cryst. Res. Technol.	22	1987	12	1557–1561

Zou Liu-juan, Zhu Guang-qu, Chen Yuan-quo, Zhang Cai-guo

Huazhong University of Science and Technology, Department of Physics, and
Wuhan Institute of Metallurgy, and
Wuhan University, Department of Physics
Wuhan, P.R. China

An Investigation on the Disordered AlNiCo8 Magnetic Steel Containing the Higher Co and Higher Ti by Means of the Positron Annihilation

In this paper, we investigate the coercive force and magnetic properties of the disordered AlNiCo8 magnetic steel containing the higher Co and higher Ti by means of the positron annihilation spectroscopy. We found that the positron annihilation lifetime τ_2 is closely related to the manufacturing technique and disorientation of this steel, and which directly affect the coercive force and magnetic behavior of the material. It is also found that the microscopic constructional characteristics in the present sample are finer than the ones of the same kind sample of U.S.A. according to an analysis of the positron average life $\bar{\tau}$.

1. Introduction

A magnetic steel, Aluminium–Nickel–Cobalt (Al–Ni–Co), is a type of permanent material which has a stable magnetic character and fine comprehensive properties. In the last decades, this material, mainly as a permanent material, has been extensively applied in many fields, such as instruments, apparates, computers, automatic-controls and electronics et al.

The magnetic properties of this alloy, Al–Ni–Co, so far as its chemical compositions are concerned essentially depend upon the contents of titanium (Ti) and Cobalt (Gould), for example increasing Ti and Co one can obtain the magnetic steel with the higher coercivity.

Institute of Wuhan Metallurgy improved the properties of this material from the chemical compositions and manufacturing technique. Since 1977, they have performed experiments and investigations which used a new heat-treatment technique of the controlled cooling speed to manufacture AlNiCo8 magnetic steel with the finer properties, they have also systematically investigated, when applying this new technique, the effects of composition (Al, Ni, Co, Cu, Ti, Niobium, et al.) variations and cooling speed on the magnetic properties.

The positron annihilation spectroscopy is a newly developed nuclear technology (Hautojärvi) in recent years. Because the positron average life time ($\bar{\tau}$) is proportional to 1/ne (where ne is the electron densities), and the crystal structures, configuration, and various kinds of defects are affected by these densities therefore one can find a particular information about the defect densities, configurations et al. in alloys by means of measuring $\bar{\tau}$, and further studies, the variations for coercive force and magnetic properties due to the densities varying.

[1]) Contributed paper presented during the European Meeting on Positron Studies of Defects, March 23–27, 1987, Wernigerode, G.D.R.

In the present paper, we investigate the newly trial-produced magnetic steel, disordered orientation AlNiCo8, containing the higher Co and higher Ti with the help of a positron annihilation technique, and also with the metallographic examination for comparison.

2. Experimental methods

The samples were smelted in a high frequency induction furnace (60 kW) and the casting alloy with the identical crystal axis was obtained after casting. The solution heat treatment for the samples were first sustained at 1250 °C, then cooled down at different blowing speeds and treated in a magnetic field at constant temperature, finally the three-stages back fire treatments were performed. The bar casting samples were cut into circle slices with 1 mm thick and diameter 15 mm, then smoothed.

Our positron annihilation instrument has two ORTEC-583 type differential constant fraction timing discriminators as energy chooser. We used radiological source Co^{60} to gain the time-resolved ratio (280 ps) of the system under the condition of Na^{22} energy window. The intensity of the positron source of Na^{22} was about 5 μc, and the carrier was a Mylar film. The source and sample formed a 'sandwich', the measurements were done at room temperature (20 ± 1 °C), the total countings of the life spectrum were 8×10^5. A simplified POSITRONFIT EXTENT program was used to analyze the positron lifetime spectrum, and to fit the three lifetimes, τ_1, τ_2, τ_3, and corresponding intensity I_1, I_2, and I_3. A longer lifetime, τ_3, is due to the positron annihilating in the source and at the surface of samples. The first and second component, τ_1 and τ_2, are taken as parameters of the positron annihilating in samples.

According to the model of three states capturing, we obtain the positron average life

$$\bar{\tau} = \sum_{i=1}^{3} \tau_i I_i \,.$$

The compositions and contents of AlNiCo8 used in our experiments are listed in Table 1.

Table 1
The compositions and contents of AlNiCo8 containing the higher Co and higher Ti

composition	Ni	Al	Co	Ti	Cu	C	S	Fe
disordered content (%)	14	7	40	7	3.00	0.037	0.159	28.8
ordered content (%)	13.5	7.6	34.5	5.7	3.5			35.2

3. Analysis and discussion

The obtained coercive force (H_c), magnetic properties $(BH)_m$, remanence (B_r) and positron annihilation lifetime data for our samples are indicated in Table 2.

1. It can be seen from Table 2 that the positron annihilation lifetime τ_2 is increased with the increase of cooling speed when casting; the positron average lifetime τ is approximately identical each other for the same group samples under the identical test conditions.

The positron annihilation lifetime values (τ_2) are distributed over 320 ~ 352 ps, for A, B group samples with the identical technological process. But, the sample $1B_2$-13-1 in A group seems a particular one, which is cut from the top end of the

Table 2
Positron life spectrum data of a disordered orientation AlNiCo8 containing the higher Co and higher Ti

	Sample No		B_r	H_c	$(BH)_m$	τ_1/ps	I_1/%	τ_2/ps	I_2/%	τ_3/ps	I_3/%	$\bar{\tau}$/ps
A	1th test	$1B_2$-13-1	7101	1383	3.46	107	74	298	20.6	1777	5.4	237*
		$1B_2$-14-2	6555	1338	2.82	117	77,3	325	17.2	1817	5.4	247
		$1B_2$-15-1	7429	1805	4.59	121	81.3	352	13.6	1927	5.1	247
		$1B_2$-15-2	6992	1585	3.69	114	76.5	320	18.1	1860	5.4	247
B	2th test	18-2E-30	9420	1655	6.8	121	77	339	18	1855	5	247
C	test for fast cooling	$1B_2$-21-1	8501	1897	5.81	125	80.1	365	15	1854	4.9	243
		$1B_2$-22-2	7756	1906	5.54	112	76.7	327	17.8	1782	5.5	240
		$1B_2$-24-1	7647	1924	5.58	119	80.4	370	14.4	1850	5.2	243
		sample of U.S.A.	9300	1565	6.28	114	75.9	343	18.4	1839	5.7	254
		86-4-0 orientation No. 8	11100	1480	10.90	110	75.9	303	18.8	1696	5.3	231
D		18-24-30	casting state			113	77.7	327	16.8	1809	5.4	241

casting bar, τ_2 and $\bar{\tau}$ are obviously lower than that of other samples. We consider it is due to the casting non-homogeneity when manufacturing the samples.

For C group samples, the positron life values (τ_2) are distributed over 327 ~ 370 ps. And the cooling speed for C group is more quickly than that for A and B.

For D group sample, the τ_2 and $\bar{\tau}$ are 327 ps and 241 ps, respectively, because of the different casting.

For A group samples, for example $1B_2$-14-2, whose cooling speed was slower, its precipitations have not precipitated much and appear to be non-homogeneous, thus the $(BH)_m$ is lower and $\tau_2 = 325$ ps. For C group samples, for example $1B_2$-21-1, the cooling speed of which is quicker, so the precipitations are more than ones of A group and even and to a greater dispersed degree, and the $(BH)_m$ is higher, $\tau_2 = 365$ ps. And those are verified by the metallographs shown in Figures 1 and 2.

A relationship between the positron life τ_2 and coercive force H_c is given in Figure 3: The curve in range I represents the samples, the cooling speed of which is slower, and in range II, cooling speed is higher. The total tendency is that the coercive force

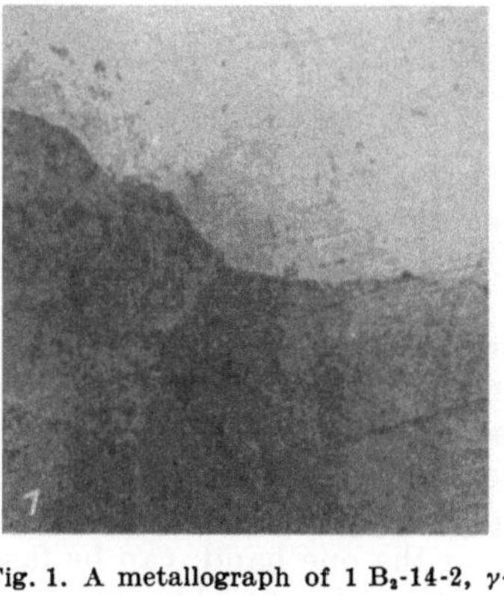

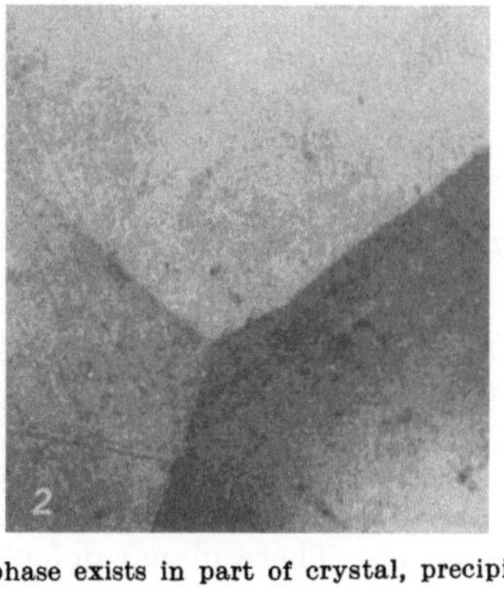

Fig. 1. A metallograph of 1 B_2-14-2, γ-phase exists in part of crystal, precipitations are not many and appear to be non-homogeneous, $(BH)_m = 2.82$, $\tau_2 = 325$ ps

Fig. 2. A metallograph of 1 B_2-21-1, precipitations are more than ones in Figure 1 and even $(BH)_m = 5.81$, $\tau_2 = 365$ ps is increased with H_c increasing, the τ_2 is higher in II than in I, and τ_2 is increased with the cooling speed increasing.

2. The orientation of magnetic materials considerably determines the positron average lifetime $\bar{\tau}$. It can be shown from Figure 5 that the curve, for our A, B, C group samples which are disordered orientation, falls above the lower curve (I) of an orientation sample No. 8, $\bar{\tau} = 231$ ps. For a sample of orientation No. 8 the manufacturing

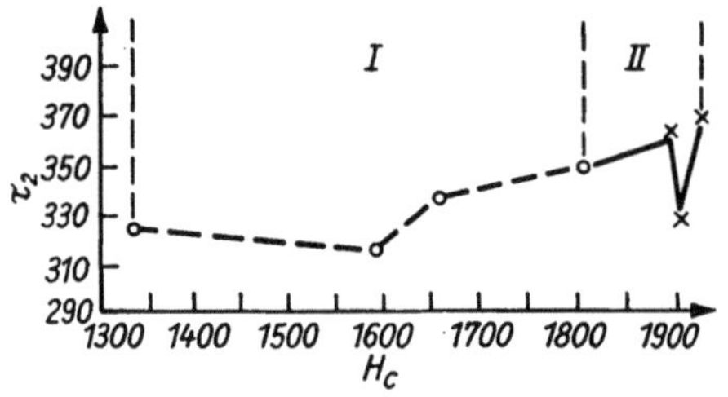

Fig. 3. Positron lifetime τ_2 and coercivity H_c

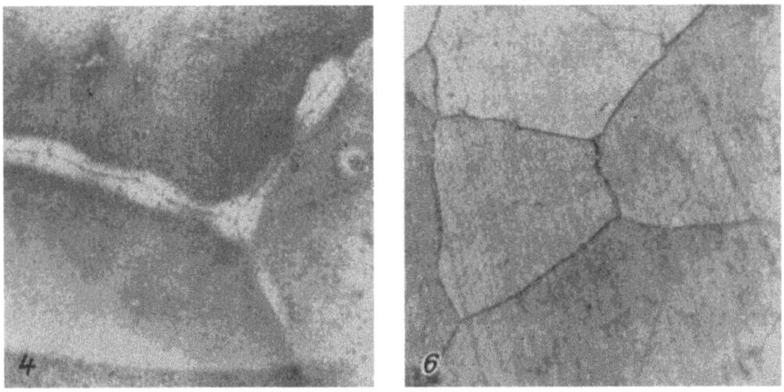

Fig. 4. A metallograph for the sample of the orientation No. 8, there is a little of feather-like γ-phase along the crystal boundary, but not precipitations

Fig. 6. A metallograph of U.S.A. sample, precipitations are thin and small, dispersed, and the size of crystal grains is even

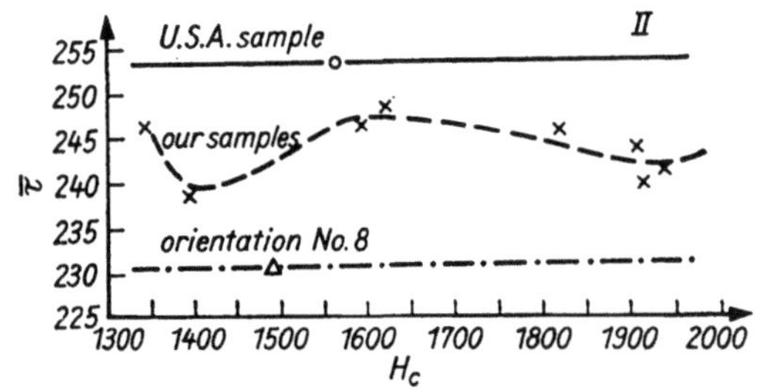

Fig. 5. Positron average lifetime $\bar{\tau}$ and coercivity H_c (lower curve: I)

technique is as follows: after cooling by air the sample is put into a magnetic field and decomposed by spinodal (Xu Ke Jian et al.). Then the magnetic phase is changed, $\alpha \to \alpha + \alpha'$. For disordered orientation samples cooled by blowing, the metallographical structures, besides the grains, present much of thickly dotted, precipitated phase, which are increased as the cooling speed increasing and so do the coercivity, τ_2 and planar defects. These planar defects, on the one hand, can be fastened up the domain wall to increase the coercivity, and on the other hand to increase the positron life τ_2. Therefore the total effects $\bar{\tau}$ of defects for the disordered orientation samples, A, B and C group, are greater than that for the orientation No. 8 sample. And there is no precipitated phase in the magnetic steel of the orientation No. 8, so the coercivity is lower (see Fig. 4 and compare with Figs. 1 and 2).

3. We compare the our three group samples (A, B and C) with the U.S.A. samples ($\bar{\tau} = 254$ ps), find our samples $\bar{\tau}$ is less than U.S.A.

The point defects, such as vacancies, impurities, vacancy clusters et al., and line defects as dislocations et al. and planar defects as crystal boundary et al. existing in metals and alloys can easily capture the positrons to make the positron lifetime increasing. The average lifetime $\bar{\tau}$ reflects the prehensive characters of microscopic structure defects, and the finer or worse for the material properties are directly affected by the quantities of various defects, and it is also verified by experiments on the macroscopic mechanical property. So, these prove the characters of our samples (A, B, and C group) are finer than U.S.A. (see Fig. 6 for comparison) in the view of two aspects, $\bar{\tau}$ and macroscopic mechanical properties.

References

GOULD, J. E.: Cobalt monograph series, Brussels 1971

HAUTOJÄRVI, P. (ed.): Positrons in Solids, Berlin 1979.

XU KE JIAN, ZHU GUANG-QU, LIU SHAN-YI.: A permanent-magnetic alloy of AlNiCo8 and spinodal decomposed crystal constructures, to be published

(Received June 16, 1987)

Authors' addresses:

ZOU LIU-JUAN
Huazhong University of Science and Technology
Department of Physics
Wuhan, P.R. China

ZHU GUANG-QU, CHEN YUAN-QUO
Wuhan Institute of Metallurgy, Department of Physics
Wuhan, P.R. China

ZHANG CAI-GUO
Wuhan University, Department of Physics
Wuhan, P.R. China

Book Review

A. Naiman, R. Rosenfeld, G. Zirkel

Understanding Statistics

International Editions, Mathematics and Statistics Series

McGraw-Hill Book Company, New York 1986; XI, 355 pages, third edition. ISBN 0-07-Y66605-9. DM 54.00

This is an unusual book on statistics. It presents sampling and hypothesis testing without mathematical theory in the same way as there are language textbooks foregoing grammatical rules. Nevertheless, it is always mathematically accurate in terminology, symbols, and formulas. At the end of each chapter you can find a vocabulary which lists the newly introduced terms ranging from median and estimate of standard deviation to Mann-Whitney U statistic. But in contrast to most textbooks on statistics for practice (cf. Cryst. Res. Technol. **21**, 1462 (1986), no new simplified definitions or auxiliary constructions are given. Topics are introduced informally by questions and examples leading naturally to the development of pertinent ideas.

The aim of this book is to show students with little interest and background in mathematics how to use statistics. No preference is given to a special field; examples refer to real-world problems and students' lives, sometimes with a humorous touch.

It is the authors' great merit that they use precise mathematical terminology without encumbering the reader with its mathematical foundations and technical details. Nevertheless, they succeed in covering all aspects that are relevant to practical application.

H. J. Girlich

Cryst. Res. Technol.	22	1987	12	1563–1567

G. AMARENDRA, B. VISWANATHAN, K. P. GOPINATHAN

Materials Science Laboratory, Indira Gandhi Centre for Atomic Research
Kalpallkkam, India

Positron Lifetime Study of Helium Clustering in Ni and Ni-1 at.% Ti[1])

Helium clustering and bubble growth have been studied in Ni and Ni-1 at % Ti through positron lifetime measurements. Helium pressure in the bubble as well as bubble concentration have been estimated and compared for the cases of Ni and Ni-1 at % Ti. The results seem to indicate enhanced helium trapping due to substitutional Ti impurity in the host Ni.

1. Introduction

Helium is practically insoluble in metals. The behaviour of helium in metals and alloys has been a topic of both basic and technological interest in recent times (ULLMAIER). Positron annihilation spectroscopy (PAS) has been successfully addressed to questions on helium-vacancy interaction, nucleation, growth and stability of helium bubbles (VISWANATHAN et al. 1983, 1985; KÖGEL et al.; HANSEN et al.). The merit of PAS for the study of helium in metals is two-fold: (a) the technique has proven ability to distinguish clearly between helium-free and helium decorated vacancy clusters, (b) the method employs bulk samples and therefore it is not complicated by helium loss to surface in the determination of helium pressure in the bubble. In this paper, positron lifetime measurements carried out in pure nickel and nickel containing 1 at.% Ti are reported. The present study is specifically addressed to the effect of Ti-alloying on bubble growth in nickel.

2. Experimental details

2.1. Helium implantation

Well annealed polycrystalline samples of nickel (5 N purity) and dilute alloy of Ni-1 at.% Ti were implanted in vacuum with 40 MeV α-particles from a cyclotron. Uniform distribution of helium over a depth of 200 μm from the sample surface was achieved by continuously degrading the alpha beam energy from 40 MeV to about 1 MeV. The implantation arrangement used is shown in Figure 1. The present implantation procedure ensured strong overlap of the helium profiles with the positron profile, thereby enhancing the strength of positron annihilation signals from helium sites. The implanted helium doses were 250 appm and 150 appm respectively in Ni and Ni-1 at.% Ti. The irradiation temperature was 450 ± 10 K.

2.2. Positron lifetime measurements

Positron lifetime spectra were measured with a fast-fast coincidence spectrometer having a time resolution of 300 ps (FWHM). The spectra with adequate statistics were analysed using the program POSITRONFIT (KIRKEGAARD et al.). Because of the conventional sample-source-sample sandwich assembly used, implantation was done on both the samples under identical conditions. The positron source used was 15 μCi ^{22}Na evaporated on 1.2 μm

[1]) Contributed paper presented during the European Meeting on Positron Studies of Defects, March 23–27, 1987, Wernigerode, G.D.R.

thick Ni foil. Source correction and spectrometer time resolution were determined from spectra of known lifetime using the program RESOLUTION (VAN DER KOLK et al.). Post-implantation isochronal annealing was performed in high vacuum between 300 to 1200 K. Positron lifetime measurements were made at room temperature after each annealing step (annealing time: 45 min for each step).

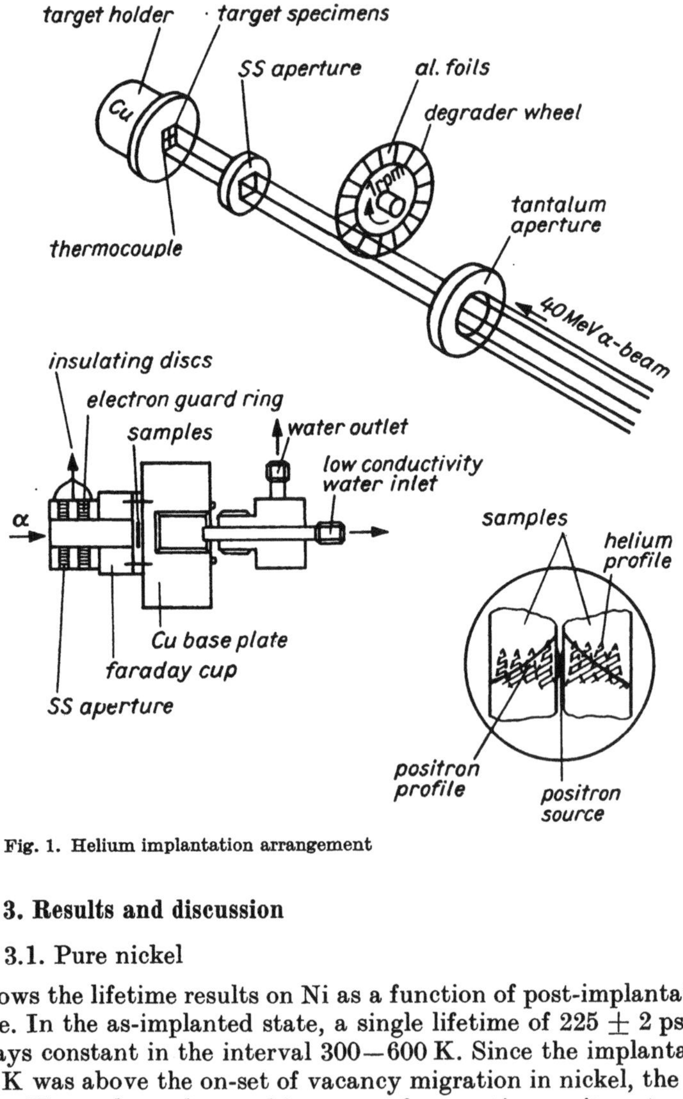

Fig. 1. Helium implantation arrangement

3. Results and discussion

3.1. Pure nickel

Figure 2 shows the lifetime results on Ni as a function of post-implantation annealing temperature. In the as-implanted state, a single lifetime of 225 ± 2 ps was observed and this stays constant in the interval 300—600 K. Since the implantation temperature of 450 K was above the on-set of vacancy migration in nickel, the lifetime behaviour till 600 K may be understood in terms of saturation positron trapping at small vacancy clusters. Above 600 K, two lifetime components are clearly resolved as seen in the figure. In the interval 600—800 K, the longer lifetime τ_2 decreases from 225 ps to 190 ps and tends to increase again. The corresponding intensity I_2 increases to a maximum before decreasing. This stage marks the re-arrangement of defect structure, wherein the undecorated vacancy clusters evolve into helium decorated vacancy clusters. This results in nucleated microbubbles of helium. The present data in this stage are in accordance with the earlier results reported in Ni (KÖGEL et al.).

Beyond 800 K, the lifetime τ_2 increases rather sharply with a corresponding reduction of its intensity I_2. At still higher temperatures, τ_2 saturates at a value of 425 ps and I_2 stabilises to a constant value of $\sim 8\%$. This may be explained by the growth of helium-vacancy agglomerates and eventual relaxation to stable equilibrium bubbles. The latter process is aided by thermal vacancy condensation. The behaviour of τ_1 and

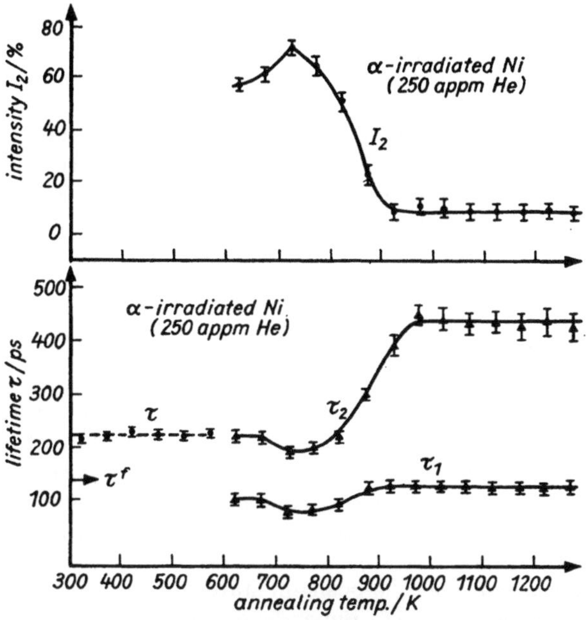

Fig. 2. Lifetime τ and intensity I_2 vs isochronal annealing temperature for α-irradiated Ni (τ is the single lifetime with 100 % intensity)

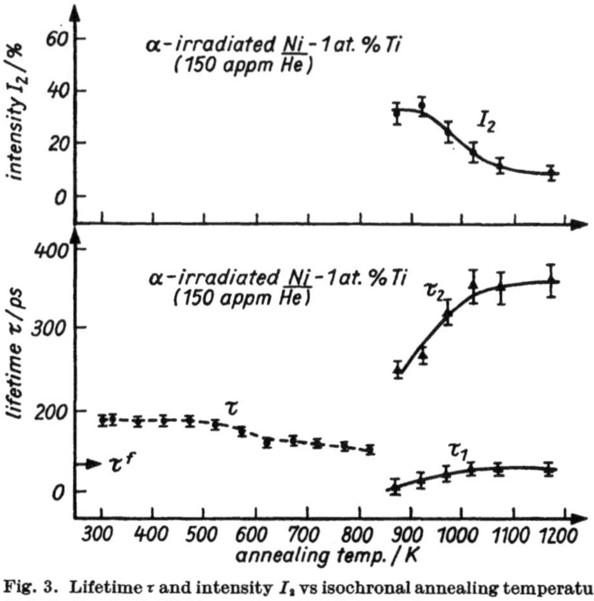

Fig. 3. Lifetime τ and intensity I_2 vs isochronal annealing temperature for α-irradiated Ni-1 at. % Ti (τ is the single lifetime with 100 % intensity)

I_1 are in accordance with the expected behaviour of the bulk state on the basis of the two-state trapping model.

3.2. Ni-1 at.% Ti

The annealing curve for the recovery of helium implanted Ni–Ti is shown in Figure 3. In the as-implanted state of the dilute alloy, the observed single lifetime is 190 ps which is smaller than the corresponding value for pure Ni. This reduction in lifetime is explained as due to the clustering of vacancies bound to Ti. Between 300—820 K only a single lifetime is observed. However, in this range the lifetime decreases, indicating an annealing step around 600 K. This observation is attributed to helium decoration of Ti-bound vacancy clusters. Beyond 820 K, two lifetimes are resolved. The longer lifetime τ_2 increases and tends to saturate at a value of 350 ps accompanied by a reduction and eventual levelling off in the corresponding intensity I_2. This behaviour, similar to that in pure nickel, is indicative of bubble growth and stability of equilibrium bubbles.

3.3. Comparison of results between Ni and Ni–Ti

The lifetime results in the growth stage at high temperatures have been analysed as discussed below. In helium-free voids, it is well known that the positron lifetime increases with size. For large voids, the lifetime becomes characteristic of surface-bound positron state and is independent of void size. On the other hand, in the case of bubbles above a certain size, the positron lifetime is controlled by the helium atom density whereas the lifetime in smaller bubbles is also limited by the bubble size (KÖGEL et al.; VISWANATHAN et al. 1985; HANSEN et al.).

On this assumption let us consider the positron state to be in the interior of a large bubble. Then, the annihilation rate λ_{He} depends on the He density n_{He} in a way similar to that found in studies on compressed He (HANSEN et al.)

$$n_{He} = \alpha \cdot \lambda_{He}\,, \tag{1}$$

where $\alpha = 3.71 \times 10^4$ ps nm^{-3}. n_{He} is obtained using (1) by equating the measured bubble annihilation rate τ_2^{-1} to λ_{He}. From n_{He}, the corresponding helium pressure at a given temperature T is deduced using the equation of state by TRINKAUS. The deduced helium pressure is equated to the equilibrium value given by

$$P = 2\gamma/r_B\,, \tag{2}$$

where γ is the surface free-energy, and r_B is the bubble radius. Using the known value of P, r_B is obtained using (2). The concentration of helium bubbles c_B is then computed using

$$K_B = \mu_B \cdot c_B\,, \tag{3}$$

where K_B and μ_B are the total and specific positron trapping rates respectively. K_B is deduced from the measured annihilation parameters using the well known two-state trapping model expressions. Knowing K_B and using the value of μ_B reported by HANSEN et al., c_B is estimated. As a check of consistency, the total concentration of helium trapped in bubbles has been computed using n_{He}, r_B, and C_B. This compares satisfactorily with the implanted helium concentrations within the uncertainty of the model scheme used. All the deduced bubble parameters are tabulated below for Ni and Ni–Ti at temperatures of 920 K and 1170 K.

As seen from the Table 1, at a given temperature, the helium pressure in the bubble for the case of Ni–Ti is found higher than that for pure Ni. The bubble radius

Table 1
Parameters characteristing bubble growth derived from experimental annihilation parameters

Sample	annealing Temp. K	measured τ_2/ps	n_{He} nm^{-3}	P GPa	r_B Å	c_B ppm
Ni	920	390	95	8	4.7	1.5
	1170	425	87	6	6.3	0.9
Ni–Ti	920	315	118	12	3.1	3.0
	1170	350	105	10	4.0	1.3

is smaller for the dilute alloy whereas the bubble concentration is higher by nearly a factor of 2 as compared to Ni. The latter result has been found even for a smaller initial helium concentration in the dilute alloy. Increased helium pressure and a reduced bubble radius suggest that bubbles in the dilute alloy are less relaxed with stronger helium packing. This effect may be understood qualitatively by invoking the idea of enhanced helium trapping due to substitutional impurities, as is evident from earlier reported desorption studies (VAN DER KOLK et al.). It may be pointed out that Ti is an oversized substitutional impurity with smaller electron density in the host nickel. Such substitutional impurity sites might act as favourable trapping centres for helium agglomeration. Further work is in progress with different Ti concentrations for a detailed understanding of the present results.

The authors thank the staff of VEC Centre, Calcutta for their co-operation.

References

HANSEN, H. E., RAJAINMAKI, H., TALFA, R., BENTZON, M. D., NIEMINEN, R. M., PETERSEN, K.: J. Phys. F **15**, 1 (1985)
KIRKEGAARD, P., ELDRUP, M., MOGENSEN, O. E., PEDERSEN, N. J.: Computer Phys. Commun. **23**, 307 (1981)
KÖGEL, G., QIN-MIN FAN, SPERR, P., TRIFTSHÄUSER, W., VISWANATHAN, B.: J. Nucl. Mater. **127**, 125 (1985)
TRINKAUS, H.: Rad. Effects **78**, 189 (1983)
ULLMAIER, H.: Rad. Effects **78**, 1 (1983)
VAN DER KOLK, G. J., VAN VEEN, A., CASPERS, L. M., DE MOSSEN, Th. M.: J. Nucl. Mater. **127**, 56 (1985)
VISWANATHAN, B., AMARENDRA, G., GOPINATHAN, K. P.: in: Positron Annihilation, eds. P. C. JAIN, R. M. SINGRU and K. P. GOPINATHAN, World Scientific Publishing Co., Singapore, 1985, p. 862
VISWANATHAN, B., TRIFTSHÄUSER, W., KÖGEL, G.: Rad. Effects **78**, 231 (1983)

(Received June 16, 1987)

Authors' address:

G. AMARENDRA, B. VISWANATHAN, K. P. GOPINATHAN
Materials Science Laboratory
Indira Gandhi Centre for Atomic Research
Kalpakkam 603102, India

Book Review

ARCHIMEDES

Über schwimmende Körper und die Sandzahl

Ostwalds Klassiker: 213, Reprint der Erstausg. 1925

Akad. Verlagsges. Geest und Portig, Leipzig 1987, 82 S., 37 Abb., M 11.—, ISBN 3-321-00020-2

Das gegenwärtig zu beobachtende Unternehmen, OSTWALDS Klassiker der exakten Wissenschaften nach zwei Generationen in Reprints neu herauszubringen, ist nicht minder verdienstvoll als die Erstausgabe. So kann der Leser anhand des vorliegenden Bändchens wieder die originale Herleitung der hydrostatischen Grundgesetze kennenlernen. Danach kann er die ARCHIMEDISCHE Theorie der stabilen und instabilen Gleichgewichtslagen eines schwimmenden, senkrecht zu seiner Achse geschnittenen Rotationsparaboloidsegments studieren. Diese erstaunliche Theorie ist trotz der relativ freien Übersetzung nicht ohne Mühe zu verstehen und nimmt sich selbst in der vom Herausgeber angefügten modernen Version langwierig aus.

In der zweiten Abhandlung entwickelt ARCHIMEDES eine Methode der Benennung extrem hoher Zahlen und führt sie bis 10^A, mit $A = 64 \cdot 10^{16}$ aus. Verglichen damit ist die „Sandzahl", die Zahl der Sandkörner, die den Kosmos des ARISTARCH ausfüllen würden, klein, nämlich 10^{63}. Kenntnis von der heliozentrischen Theorie dieses Astronomen haben wir übrigens nur aufgrund der vorliegenden Schrift des ARCHIMEDES, welche auch einen kurzen Einblick in die Praxis der antiken Astronomie gewährt.

H.-G. SCHÖPF

Cryst. Res. Technol. | 22 | 1987 | 12 | 1569–1573

A. Z. Ilyasov

Institute of Electronics, Uzbek Academy of Sciences, Tashkent, USSR

Positron Trapping by Radiation Defects in Metals[1])

Методом измерения угловой корреляции аннигиляционных гамма-квантов исследованы процессы образования дефектов в молибдене, облученном протонами с энергией 30 МэВ флюэнсом 4,5 · 10^{16} см$^{-2}$. Наблюдались 4 стадии отжига в интервале температур 300—1200 К.

Исследованы также дозные зависимости параметров кривых угловой корреляции гамма-квантов, испускаемых при аннигиляции позитронов в образцах железа и тантала, облученных ионами азота с энергией 35 МэВ.

1. Introduction

Rapidly developing nuclear energetics puts forward new strict requirements to construction materials. In this connection radioresistant materials are being searched for. Positron annihilation method provides a valuable information on the process of radiation defect formation in metals and alloys. For instance, energies of vacancy formation have been measured actually in all metals by this method. The most complete set of these data is given in the review by Doyama. These data are very important for the creation of a material for the first wall of a future fusion reactor.

For this purpose it is necessary to study the behaviour of materials-claimants (first of all these are refractory metals such as tungsten and molybdenum and metals of the iron group) under conditions being close to real ones. As the sources of fast neutrons with the required intensity are not available yet the investigations are mostly performed in charged particle accelerators with whose help it is possible to simulate the effect of neutrons on metals (Kaletta) and to study the process of metal blistering during their irradiation by α-particles and protons. The positron annihilation method permits to study pore formation at the earliest stages when other methods are not sensitive enough.

The investigation results of defect formation in molybdenum irradiated by accelerated protons and in tantalum and iron irradiated by nitrogen ions are presented below.

2. Experimental

To study radiation defects we used the method of angular correlation of annihilation radiation (ACAR). ACAR measurements were performed in the installation with a point-linear geometry and angular resolution of 1.1 mrad (the arm length was 2 m). The sample and the source were positioned in the vacuum chamber.

The chamber design provided a precise sample loading and its heating up to 1500 °C. While making the temperature measurements the positron source was cooled by flowing

---

[1]) Contributed paper presented during the European Meeting on P Positron Studies of Defects, March 23–27, 1987, Wernigerode, G.D.R.

water. ^{22}Na isotope with the activity of 10 mCi on a glass substrate was used as a positron source.

3. Experimental results and their discussion

Molybdenum.

Due to the composition of such properties as a high melting point, a low evaporation rate, a small cross-section capture of fast neutrons and a low guided activity molybdenum can be considered as one of the most promising materials for the first wall of a fusion reactor.

Molybdenum samples with the thickness of 0.2 mm were formed as a pile with the height of 1.6 mm and irradiated by protons accelerated in a cyclotron up to the energy of 30 MeV and fluence of 4.5×10^{16} cm^{-2}. During irradiation the samples were cooled by flowing water at the pressure of 4 atm. In this case first samples were heated to the temperature that did not exceed 200 °C. The measurement results of ACAR are summarized in Table 1.

Table 1
ACAR curve parameters for molybdenum samples irradiated by protons with the initial energy of 30 MeV

Sample No.	count number in the maximum ACAR curve H/rel. units	FWHM/mrad	$f = \frac{N(0)}{N(\vartheta)}$
Initial	0.078 ± 0.001	12.4 ± 0.2	2.54
1.	0.095 ± 0.001	9.4 ± 0.2	3.86
2.	0.098 ± 0.001	8.7 ± 0.2	4.06
3.	0.096 ± 0.001	9.0 ± 0.2	3.92
4.	0.094 ± 0.001	9.6 ± 0.2	3.31
5.	0.094 ± 0.001	9.6 ± 0.2	2.82
6.	0.092 ± 0.001	9.4 ± 0.2	3.06
7.	0.095 ± 0.001	9.5 ± 0.2	3.88
8.	0.092 ± 0.001	10.4 ± 0.2	4.34
2 { 197 °C	0.094 ± 0.001	9.3 ± 0.2	3.51
2 { 667 °C	0.082 ± 0.001	11.5 ± 0.2	3.03

While discussing the results one should bear in mind that at proton energy of 30 MeV in addition to efficient formation of Frenkel pairs there can be non-elastic and other nuclear interactions (Ibragimov et al.) resulting in the formation of large disordered regions. Due to the fact that the first sample was at the temperature of 200 °C annealing of divacancies and partially of vacancies occured in it according to (Eldrup et al.). That is why it is reasonable to assume that after irradiation the first three samples have vacancy loops, clusters, loops of interstitial atoms, micropores and also the most stable complexes, i.e. an interstitial atom-impurity. In the last samples of the pile there are vacancies, divacancies, small clusters and complexes as an implanted atom-impurity, an interstitial atom-impurity and clusters of interstitial atoms.

The defect influence on the parameters of ACAR curves is displayed in the narrowing of the latter ones. In this case point defects, divacancies, etc. result in smaller changes of ACAR curves than clusters and pores.

To analyse the shape of ACAR curves we used the parameter $f = \frac{N(o)}{N(\vartheta)}$ introduced by Dekhtyar which is a ratio of count rates of coincidences in the maximum of ACAR curve ($\vartheta = 0$) and at the angle ϑ. The angle ϑ is selected according to the position of the extremum in the negative area of the differential curve which is the most sensitive to the variations in the characteristics of positron annihilation (PETERSEN et al.).

According to the data of Table 1 the behaviour of f-parameter qualitatively confirms the preliminary conclusions about the nature and distribution of defects in the samples. The parameter "f" has a minimum in the area of the sample No. 5 and two maxima in the sample No. 2 and 8.

It is natural to assume that there are no pores in the samples No. 7 and 8. This supposition is confirmed by the fact that in one ionization cycle created by 2 MeV protons the energy of 35 eV is released while the threshold energy of pore formation is 40 eV (GINFALCO, STREETMAN). Annealing of the sample No. 2 results in 4 stages in the temperature range of 20—900 °C which is in a good agreement with the results of (ELDRUP et al.).

Under the assumption that during irradiation the samples No. 7 and 8 have only vacancies and in the sample No. 2 there are vacancies and pores and taking into consideration the fact that after annealing at temperatures above 600 °C this sample has only pores then within the framework of the trapping model (WEST) and (DEKHTYAR et al.) we calculated the concentration of pores and vacancies C_p and C_v according to Eq. (1) (DEKHTYAR et al.):

$$f = \frac{f_0\lambda_f + f_v C_v \sigma_v + f_p C_p \sigma_p}{\lambda_f + C_v \sigma_v + C_p \sigma_p}, \tag{1}$$

where f_0, f_v, and f_p are annihilation parameters in a defect-free sample and in a sample saturated with vacancies and pores, σ_v and σ_p are cross-sections of positrons trapped by vacancies and pores, λ_f is the annihilation rate in a defect-free crystal. As the annealing of the sample No. 2 is similar to the annealing of Mo irradiated by electrons (ELDRUP et al.) it can be assumed that after annealing at 667 °C in the sample No. 2 there are pores with the average diameter of $2R_p = 30$ Å. Using the ratio:

$$\sigma_p = 4\pi D_+ R_p (1 + R_p \sqrt{\lambda_f / D_+}),$$

where D_+ is a diffusion coefficient of positrons in Mo and taking it as $D_+ \cong 10^{-2}\ \text{cm}^2\,\text{s}^{-1}$ (DEKHTYAR et al.) we get $\sigma_p = 2.14 \times 10^{-8}\ \text{cm}^3\,\text{s}^{-1}$. Assuming that $f_p = 15.56$ (ELDRUP et al.) and $C_v = 0$ we have C_p (667 °C) $= 1.5 \times 10^{16}\ \text{cm}^{-3}$. Considering the conditions of sample irradiation and expecting that during the following annealing the bulk of the defect region is not changed (DEKHTYAR et al.), i.e. C_v (20°) $\times$ V_v (20 °C) $+$ C_p (20 °C) $\times$ V_p (20 °C) $=$ C_p (667 °C) $\times$ V_p (667 °C) (here V_v and V_p are the bulks of a vacancy and a pore respectively) and that C_v (20°C) $\gg$ C_p (20 °C) we get the estimation of the initial concentration of point defects

$$C_v\ (20\ °\text{C}) = 1.2 \times 10^{19}\ \text{cm}^{-3}.$$

On the strength of these data f_v parameter was estimated. The analysis of the measurement results within the trapping model permitted to define the concentration of defects (vacancies and pores) C_v and C_p for some samples. The calculation results are presented in Table 2.

As it is seen from the table in spite of rough simplifications the obtained defect concentrations are in a good agreement with the data of other works (ELDRUP et al.) and (DEKHTYAR et al.) and qualitatively confirm the common nature of defect distribution in the samples. It should be noted that the result for the sample No. 8 is distorted by hydrogen as the proton path ends exactly in it.

Table 2
Concentration of vacancies C_v and pores C_p in some molybdenum samples irradiated by protons with initial energy of 30 MeV

Sample No. concentration	5	6	7	8	2		
					20 °C	197 °C	667 °C
C_v, 10^{17} cm^{-3}	6.56	14.16	81.5	332.8	123.8	25.86	0
C_p, 10^{16} cm^{-3}	0	0	0	0	0	1.03	1.50

Iron and tantalum.

The annealed samples of Fe and Ta were irradiated by nitrogen ions accelerated in a cyclotron up to the energy of 35 MeV. As a result different types of defects and also the compounds of metal nitrides type were formed in a layer with the thickness of 30–40 μm. The variation of an electron structure occurring in this case are revealed in the adequate change of ACAR curve parameters. Table 3 presents FWHM values for Fe and Ta irradiated by different fluences of nitrogen ions.

Table 3
FWHM values of ACAR curves for iron and tantalum irradiated by different fluences of nitrogen ions with energy of 35 MeV

Sample	fluence/cm^{-2}				
	0	10^{13}	5×10^{13}	10^{14}	5×10^{14}
iron	10.1 ± 0.2	10.3 ± 0.2	10.7 ± 0.2	10.9 ± 0.2	8.5 ± 0.2
tantalum	9.7 ± 0.3	9.5 ± 0.3	9.5 ± 0.3	9.1 ± 0.3	8.1 ± 0.3

The obtained data show that beginning from the fluence of $\sim 10^{14}$ cm^{-2} one observes a narrowing of ACAR curves for iron. The similar narrowing for tantalum starts with the fluence of 5×10^{13} cm^{-2}.

4. Conclusion

It is shown in the present paper that the positron annihilation method can be successfully used for the estimation of defect concentration (vacancies, pores) formed in metals during irradiation by high-energy particles (protons, heavy ions).

The method is sensitive and quite efficient for the investigation of structural peculiarities of materials even in the cases when traditional methods of the analysis are not applicable due to their restricted resolution ability.

References

DEKHTYAR, I. Ya.: Phys. Rep. **9P**. 254 (1974)

DEKHTYAR, I. Ya., ADONKIN, V. T., RUSTAMOV, Sh. A., DEKHTYAR, M. I., SEDOV, Yu. A.: Fizika metallov i metallovedenie **58**, 292 (1979)

DOYAMA, M.: Positron annihilation. Proc. 5 Internat. Conf. Positron annihilation, eds R. R. HASIGUTI and K. FUJIWARA Sendai 1980

ELDRUP, M., MOGENSEN, O. E., EVANS, J. H.: J. Phys. H.: Metal Phys. **6**, 499 (1976)

GINFALCO, L. A., STREETMEN, J. R.: J. Phys. Chem. Solids **4**, 182 (1958)

IBRAGIMOV, Sh. Sh., CHURSIN, G. P., KOZIN, A.: VIII Internat. Conf. on Atomic Collisions in Solids, Sept. 12–23, Moscow 1977

KALETTA, D.: Proc. 1 Internat. Conf. on Ion Beam Interact. Modification of Materials, eds J. GYULAI, J. LOHNER, E. RAZTOR, Budapest 1979
PETERSEN, K., MOGENSEN, O., COTTERIL, R. M. J., HUDSON, B.: Nature **240**, 236 (1973)
WEST, R. N.: Advances in Physics **22**, 263 (1973)

(Received June 16, 1987)

Author's address:

A. Z. ILYASOV
Institute of Electronics, Uskek SSR Academy of Sciences
Academgorodok
700143 Tashkent, USSR

Book Review

D. F. Moore (ed.)

Solid State Devices 1986

Institute of Physics Conference Series Number 82

Institute of Physics Publishing Ltd., Bristol 1987, 201 pages, ISBN 0-85498-173-X, ISSN 0305-2346, £ 22.50.

The book contains twelve of the invited papers presented at the Sixteenth European Solid State Device Research Conference (ESSDERC) held at the University of Cambridge, Sept. 8—11, 1986. These papers are representative of the main topics of interest at the conference and are mainly related to the fabrication technology of devices. Eight of the contributions are devoted to problems in silicon device technology. The development of the two key power devices (power MOSFET and insulating gate transistor) is reviewed by Baliga. Aspects of the molecular beam epitaxy of silicon and their application in device technology are considered in detail by Bean, including also the growth of Ge_xSi_{1-x}/Si strained layer heterostructures. The fabrication and properties of modulation doped field effect transistors based on Ge_xSi_{1-x}/Si heterostructures are described by Kasper and Daembkes. Other problems considered are the use of dry etching in silicon device technology and a review of the present understanding of the underlying physics and chemistry of the process (Deshmukh), the recent progress made in optical and X-ray lithography in the submicron range including a detailed discussion of the general principles and technical problems (Heuberger), and a detailed experimental study of the microstructure of the Si/SiO_2 interface (Henzler). Aspects of submicron technology for megabit RAMS are discussed by Klaasen from the viewpoint of applied physics, and Mole gives a description of both the numerical techniques and physical models used in present day semiconductor device simulators including a discussion of the strengths and limitations of these tools. Four papers deal with problems related to the technology of devices based on GaAs. Metalorganic molecular beam epitaxy as a new technique for the controlled deposition of $A^{III}B^{V}$ semiconductor thin films with main emphasis on growth and doping of GaAs is described by Lüth. The present day state of material, process technology, design, test and use of monolithic microwave integrated circuits based on GaAs is described by Myers, while Fischer and Morkoc give a review of recent advances achieved in the use of lattice mismatched epitaxial technology to produce devices based on GaAs deposited onto Si substrates. Finally, Shibatomi describes the present situation in the development and practical use of passivation technologies for GaAs including both native oxides and dielectric films.

H. Neumann

Cryst. Res. Technol.	22	1987	12	1575–1579

J. PAJAK, R. PIETRZAK, R. SZATANIK, B. ROZENFELD

Institute of Physics, Pedagogical University, Opole, and
Institute of Experimental Physics, University Wrocław, Poland

Application of the Positron Annihilation Method for Investigations of the Diffusion of Helium in Metals[1])

A method which permits to determine the diffusion coefficient of helium in metals from the measurements of the time dependence of the intensity of annihilation radiation has been outlined. The diffusion of helium cold-rolled and cold-rolled and annealed nickel has been investigated. It has been found that the value of the diffusion coefficient of helium along the grain boundaries and dislocations in nickel is of the order of 10^{-13} cm² s^{-1}.

In der Arbeit wird ein Abriß einer Methode vorgestellt, welche erlaubt, die Diffusionskoeffizienten von He in Metallen auf Grund der zeitlichen Änderungen der Annihilationsstrahlung zu berechnen. Die Diffusion von He in Ni – gewalzt und ausgeglüht – wurde untersucht. Der Diffusionskoeffizient längs der Korn- und Versetzungsgrenzen wurde mit 10^{-13} m² s^{-1} bestimmt.

1. Introduction

The solubility of helium in metals is very weak (DRIESK), but due to the reactions (n, α) or directional injection of α-particles, the helium content can distinctly exceeds the equilibrium limit in many materials likewise in reactor ones. The helium atoms, dissolved in solid metals strongly interact with the vacancy-type crystal lattice defects, and under appropriate conditions can form the gas-bubles in a metal. The introduction of helium, even at small concentration, leads to the distinct changes of the physical properties of a metal. Therefore the investigations of the diffusion of helium in metals are of great importance for the reactor technology. Despite of this, the data on the diffusion of helium in metals are rather scarce.

The purpose of the present paper is to outline a method for determination of the diffusion of helium in metals (e.g. in nickel), based on the annihilation of positron trapped at crystal lattice defects. The method has been elaborated in 1976 by ROZENFELD et al. and applied for the first time in literature for determination of the diffusion coefficient of hydrogen in nickel.

2. Experiment

The sample were prepared from the high purity 6 N nickel. Two kinds of samples were used: a) the samples containing both the dislocations and vacancy-type crystal lattice defects at high concentrations by the cold rolling, b) the samples containing mainly the dislocations. The samples (b) were prepared by tempering the cold rolled nickel samples at 200 °C by 6 hours, followed by slow cooling to the ambient temperature. The helium atoms were implanted into the samples in a glow discharge in spectroscopically pure helium

---

[1]) Contributed paper presented during the European Meeting on Positron Studies of Defects, March 23–27, 1987, Wernigerode, G.D.R.

between the investigated sample (cathode) and anode made of a thin platinum wire. The implantation process was carried under pressure of 6 Tr at the current density of 4 mA cm^{-2}. The temperature of the samples during the implantation process was 350 °C. The electrical resistance of the samples was measured at the room temperature before, and after the implantation process. It has been found that the charging with helium causes an increase in the resistance by about 0.35%. We suppose that this increase can be ascribed mainly to the presence of helium in the metal, especially to the helium atoms trapped at the crystal lattice defects.

The angular correlation curves of the positron annihilation quanta were measured for the samples charged with helium $Ni(\theta)_{dHe}$ as well as for samples without helium $Ni(\theta)_d$ are presented as a relative difference curves (Fig. 1) (PAJAK, ROZENFELD). It has been

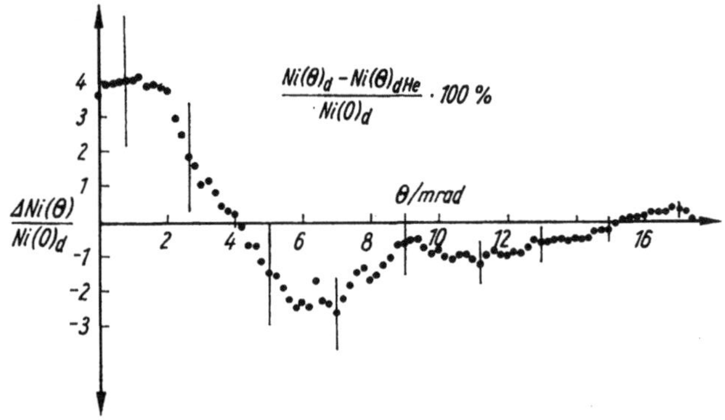

Fig. 1. Relative difference curve of the angular correlation curves for helium free and charged with helium defected by cold rolled Ni samples $Ni(0)_d$ is $Ni(\theta)_d$ for $\theta = 0$

found that the central part of curves obtained for the cold rolled nickel charged with helium is distinctly lowered with respect to the curves obtained for the cold-rolled samples without helium. According to the trapping model of annihilation process (WEST), the effect observed could be ascribed to the diminution of the relative share of the positrons annihilating being trapped at the crystal lattice defects.

The observed effect of helium on the angular correlation curves for nickel suggest that the investigations of the kinetics of trapping of the helium atoms at centres acting as positron traps enable to determine the coefficient of diffusion of helium in metals. In the next paragraph a short derivation of the method permitting to do this will be presented.

3. Results and discussion

Long, polycrystalline nickel samples of the dimensions of 50 × 10 × 2 mm³ was charged with helium in its end of the parts (ab and cd — see Fig. 2), up to concentration

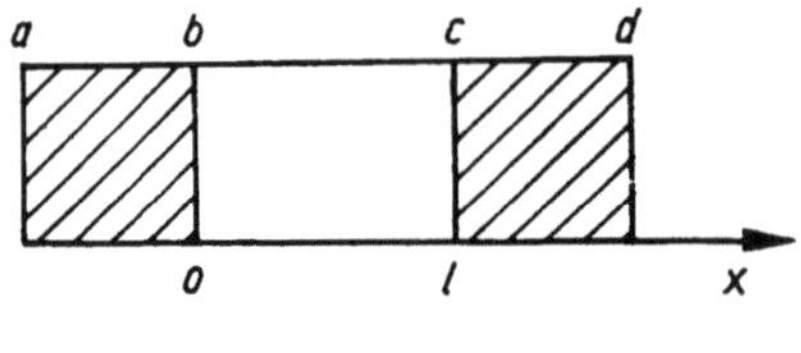

Fig. 2. A sketch of the sample charged with helium used for the measurements

not exceeding 10^{-6} at.He/at.Ni. The central part of the sample (bc) was screened during the implantation process in order to prevent the charging with helium. Due to the diffusion of helium the concentration of the unfilled positron trapping centres n_{tr}

in the central part of the sample changes according to the relation

$$n_{tr}(t) = n_{tr}(0) - n_{He}(t) , \tag{1}$$

where n_{tr} denotes the initial ($t = 0$) number of the positron traps and $n_{He}(t)$ — the number of helium atoms present in the central (bc) part of the sample at the moment t. Thus

$$dn_{tr}(t) = -dn_{He}(t) . \tag{2}$$

Basing on the model of the annihilation process described above, it can be assumed that the decrease of the number of positron traps corresponds to the decrease on the peak counting rate $dN(t)/N(0)$. Consequently,

$$\frac{dN(t)}{N(0)} = -k\, dn_{tr}(t) = k\, dn_{He}(t) . \tag{3}$$

Assuming that for $t = 0$, $n_{He}(0) = 0$ and $N(t) = N(0)$, one obtains the following solution of Eq. (3):

$$n_{He}(t) = k\,\frac{N(0) - N(t)}{N(0)} = k[1 - r(t)] , \tag{4}$$

where $r(t) = N(t)/N(0)$ and k-proportionality factor ($k < 1$) describing the probability of the annihilation of trapped positrons and the efficiency of the detection of the annihilation quanta at the applied geometry of measurements.

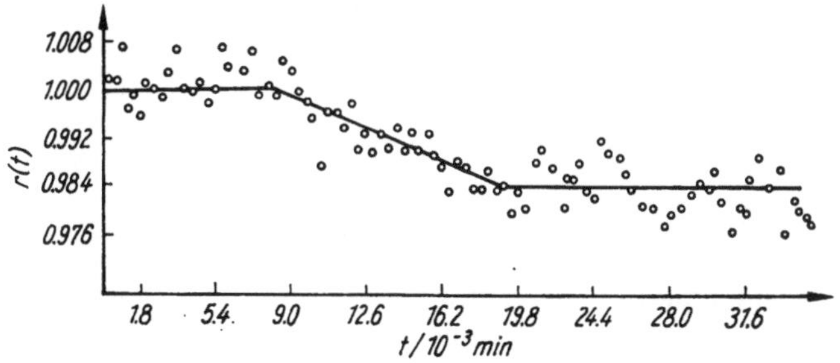

Fig. 3. Time dependence of the peak counting rate for sample a. Further explanations in text

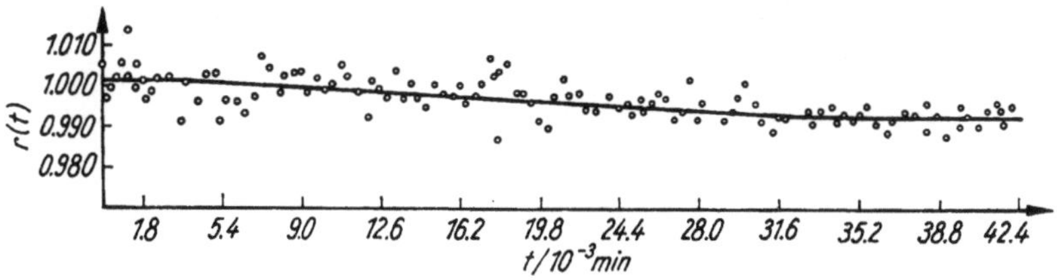

Fig. 4. Time dependence of the peak counting rate for sample b. Further explanations in text

Figures 3 and 4 represents the time dependences of the relative peak counting rate $r(t) = N(t)/N(0)$ measured integrally from the central part of the sample for the sample a, containing dislocations and non-equilibrium vacancies and for sample b, containing dislocations and equilibrium vacancies, respectively. As it can be seen, in some time interval, the relative peak counting rate decrease and then remains to be constant. If the concentration of the trapping centres $c_{tr}(0)$ in the central part (bc) of the sample is much smaller than the concentration of helium c_{oHe} in the end parts (ab and cd) ($c_{tr}(0) \ll c_{oHe}$), it can be assumed that c_{oHe} = constant.

Assuming the following initial conditions:

$$c(x, 0) = 0 \quad \text{for} \quad 0 < x < l$$

and

$$c(0, t) = c(l, t) = c_{oHe} = \text{const} .$$

we can write that number $n_{He}(t)$ of helium atoms, which diffused to bc space of the sample is equal

$$n_{He}(t) = \int_0^1 c(x, t)\, dx = c_{oHe}\, l \left\{1 - \frac{4}{\pi 4} \sum_{k=0}^{\infty} \frac{1}{(2k+1)} \exp\left[-\frac{(2k+1)^2 \pi^2}{l^2} Dt\right]\right\}, \tag{5}$$

where D is the diffusion coefficient of helium in nickel, for $t > 4.5 \times 10^{-2}\, l^2/D$ we can take into account only two first segments from equation (5) ($k = 0, 1$). We can write eq. (5) as

$$n_{He}(t) = c_{oHe}\, l \left(0.28 + \frac{7.11}{l^2} Dt\right). \tag{6}$$

Assuming that after time of the saturation t_n, number of the trapping centres in the space bc of the sample is zero, on the ground of the eq. (1) we obtain

$$n_{He}(t) = n_{tr}(0) . \tag{7}$$

As it follows from eq. (4)

$$n_{He}(t) = k[1 - r(t_r)] . \tag{8}$$

Thus, from eqs. (6)—(8) we obtain finally equation

$$r(t) = 0.72 + 0.28 r(t_n) + r(t_n) - 1] \frac{7.11}{l^2} Dt . \tag{9}$$

As it be seen the last equation describes fairly well the linear changes the peak counting rate occuring in some, limited, time interval on the curves in Figures 3 and 4. This seems to corroborate the diffusion mechanism of the process of filling the positron traps by the helium atoms.

The values of the diffusion coefficient of helium in nickel, calculated from eq. (1) using the experimental data are:

$$D^{(a)} = 2.24 \times 10^{-13}\ \text{m}^2\ \text{s}^{-1}$$

for sample a (cold rolled, not-tempered), and

$$D^{(b)} = 8.03 \times 10^{-12}\ \text{m}^2\ \text{s}^{-1} .$$

The values of the diffusion coefficient of helium in well-annealed nickel, reported by PHILIPPS and SONNENBERG, extrapolated to the room temperature, range from 8.98×10^{-16} to 7.96×10^{-18} m² s⁻¹. The value of the coefficient of diffusion of helium along the grain boundaries and dislocations in nickel, calculated by LYNE and GOODHEW equals 2.1×10^{-13} $\text{m}^2\ \text{s}^{-1}$. Thus, the values obtained in the present work suggest that the observed diffusion of helium proceeds, most probably, mainly along the dislocations and grain boundaries. The value of the diffusion coefficient for the not-tempered sample, being smaller than that for tempered sample, results, most probably, from the fact that a part of the helium atoms migrate by much slower mechanism.

4. Conclusions

From the investigations described above it follows that:
— the positron annihilation technique can be used investigations of the diffusion of helium in metals,
— in cold-rolled nickel the diffusion of helium proceeds mainly along the grain boundaries and dislocations,
— at the room temperature the value of the diffusion coefficient of helium in cold-rolled nickel is of the order $10^{-13}-10^{-12}\ m^2\ s^{-1}$.

This work has been supported by scientific program CPBP 01.09 coordinated by Institute of Physics, Jagiellon University in Cracow, Poland.

References

von den Driesk, H.: Thesis, RWTH Aachen 1980
Lyne, P. L., Goodhew, P. I.: Phil. Mag. **A48**, 965 (1983)
Pajak, J., Rozenfeld, B. (will be published)
Philipps, V., Sonnenberg, K.: J. Nucl. Mat. **107**, 271 (1982)
Rozenfeld, B., Chabik, S., Pietrzak, R., Rudzińska, W.: Fourth International Conference on Positron Annihilation, Helsingör, Denmark. 23—26 August 1976, Part 1, D11, p. 54
West, R. N.: Adv. Phys. **22**, 273 (1973)

(Received June 16, 1987)

Authors' addresses:

J. Pajak, R. Pietrzak, R. Szatanik
Institute of Physics, Pedagogical University
Oleska 8
45052 Opole, Poland

B. Rozenfeld
Institute of Experimental Physics, University of Wrocław
Wrocław, Poland

Crystal Research and Technology

Zeitschrift für experimentelle und technische Kristallographie

Edited by H. NEELS, Leipzig

Papers in Number 1, 1988

Review

Original Papers

AKADEMIE-VERLAG · BERLIN

Cryst. Res. Technol.	22	1987	12	K219–K223

Short Notes

Selina NIGLI, C. JAGADISH, G. K. CHADHA

Department of Physics and Astrophysics, and Sri Venketeswara College
University of Delhi, Delhi, India

X-Ray, Optical and Electrical Study of Melt-grown CdI_2 Single Crystals

CdI_2 has been purified using the zone-refining technique by giving 20 passes. Single crystals grown from the starting material showed arcing in the X-ray oscillation photographs whereas it is absent in the crystals grown from the zone refined material. These crystals are found to be of the 4 H type. DC conductivity studies were made on these crystals at various stages of purification. The surface morphology of these crystals have been studied using an optical microscope.

1. Introduction

The photochemical property of CdI_2 has led to its application in micro-recording, holography, printed circuits, and digital storage systems (TUBBS 1974; BUROFF et al.). The production of high quality single crystals is of prime importance in such applications. Holographic gratings require very pure layers of excellent optical quality and thickness of the order of 0.1 μm. Crystals have been grown by various techniques and their optical, photochemical, electrical and various other properties widely studied. (BEST; TUBBS 1968; GOTO, UETA; WRIGHT, TUBBS; GAHWILLER, HARBECKE; JONES et al.; LYSKOVICH et al.) Compared to other methods there has been relatively less amount of work done on crystals grown from the melt. The present investigation is an attempt to study, in depth, the effect of impurities on the X-ray, optical and electrical properties.

2. Experimental

The details of the experimental set-up used for the zone refining and single crystal growth has been described elsewhere (NIGLI 1986; NIGLI et al. (a)). Analar grade CdI_2 was used as the starting material. Though CdI_2 is the most anhydrous of metal halides it does contain small amounts of water which is responsible for the sticking of the ingot to the container due to oxide formation (BROWN; CLARK, MITCHELL). The sticking was eliminated by continuous charge container replacements. CdI_2 does exhibit evaporation and decomposition in the molten state (GMELIN). However, neither of these processes proceeds at a rate sufficient to interfere with the zone melting operation. Pure argon was used as an inert atmosphere to prevent oxidation and supress evaporation of the molten material. CdI_2 is found to expand upon melting, causing matter transport in a direction opposite to that of zone travel. This matter transport was eliminated by appropriate tilting of the container (SHAH). At different stages of the zone refining, the ingot was removed, cropped (SHAH) and sectioned along its length, in dimensions of $4 \times 3 \times 2$ mm for physical estimation of the purification process. Sections of the crystal ingot were also taken for X-ray and optical study.

Single crystal growth was the last operation of the zone refining processes. A single crystal seed was placed at the initial end of the container carrying the 20 pass material. The molten zone was initiated at a position which ensured that a portion of the seed melts into the molten zone, thus eliminating dislocations at the initial join of the crystal.

Figure 1 shows the temperature profile of the crystal growth.

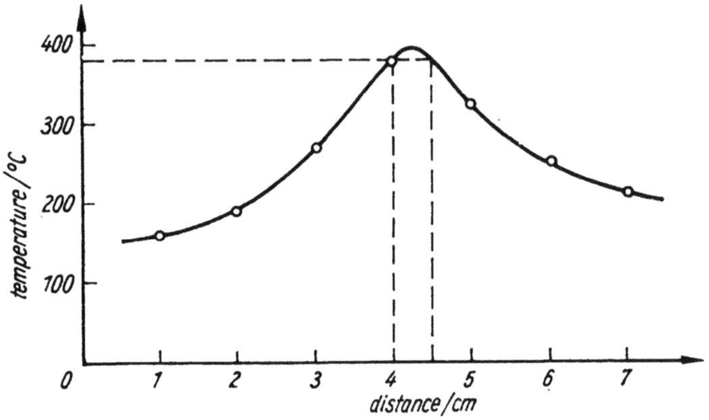

Fig. 1. Temperature profile for crystal growth of CdI_2

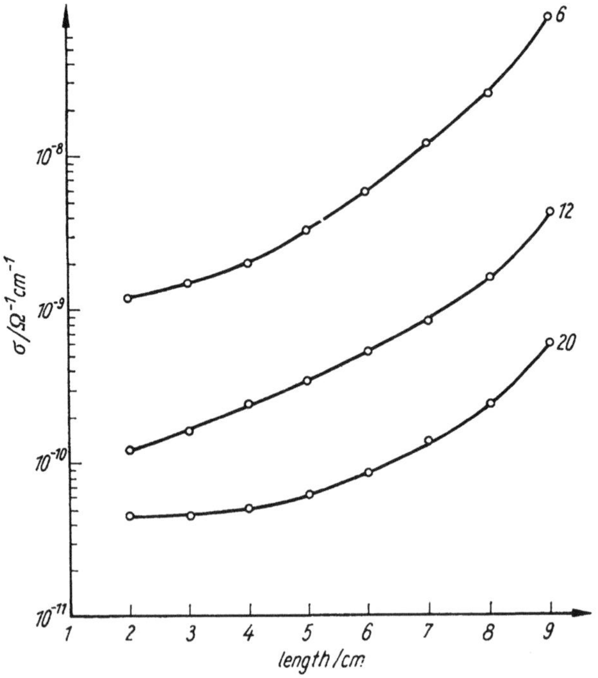

Fig. 2. Variation of DC conductivity (σ) along the length (l) of the ingot (log σ versus l), 6, 12, and 20 passes

Results

Electrical conductivity measurements of sections of the ingot along its length for different stages of purification were made. Figure 2 shows variation of dc conductivity along the length of the ingot. The slope of the curve indicates that purification is taking place. The decrease in conductivity at the final ends for increasing number of zone passes is a result of repeated croppings. The general increase in conductivity in the direction in which zoning is taking place shows that segregation occurs in the direction of zoning, indicating $k < 1$ for CdI_2.

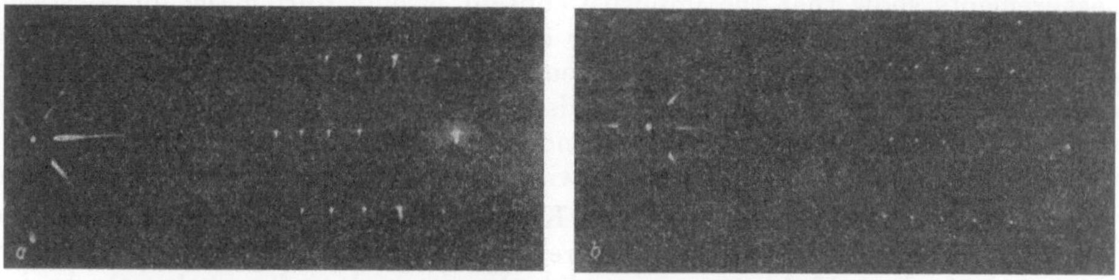

Fig. 3. *a*-axis X-ray oscillation photographs of CdI_2: (a) single pass, (b) 20 passes

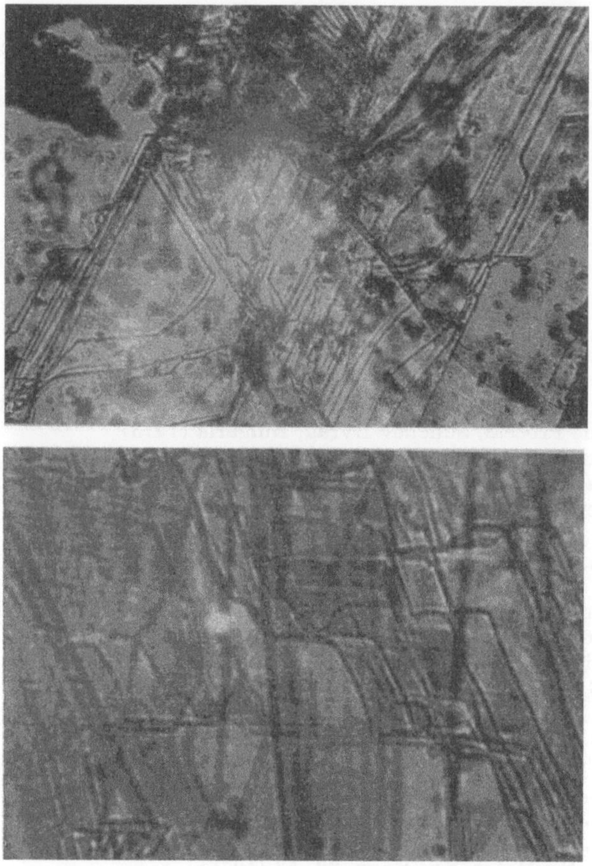

Fig. 4. Microphotographs showing surface morphology of different parts of the CdI_2 single crystals

X-ray oscillation photographs were taken at different stages of purification for a number of sections of the crystal ingot using CuKα radiation. In the crystals obtained during the initial stages of purification the spots on the X-ray oscillation photograph were found to be extended while they were sharp spots in the later stages as seen in Figures 3a and b, respectively. The presence of impurities in the starting material is responsible for the creation of edge dislocations in the crystal which in turn are responsible for the extension of spots. In the relatively pure material ($<$ 12 passes) the number of edge dislocations decreases and the extension in the spots disappears.

All the crystals examined showed only the 4 H polytype. Electrical conductivity measurements show that the impurity content is high during the initial stages of purification. If these impurities had any effect on the polytypism in CdI_2, polytypes other than the basic 4 H should have occurred in the crystals obtained from material with a fewer number of zone passes. This has, however, not been observed. Specific impurities have also been introduced and its effect on the polytypism in CdI_2 has been studied (NIGLI 1986). The results will be published elsewhere (NIGLI et al. (b)).

The faulted matrix theory (PANDEY, KRISHNA 1975a, b, c, 1976), the most recent theory put forward to explain the occurrence of higher polytypes, requires the basic matrix to be faulted. If this theory is valid, then it can be inferred that the basic 4 H matrix of CdI_2 remains undisturbed under the prevailing conditions of growth in the present investigation.

Figure 4 shows microphotographs of the (001) face of the CdI_2 single crystals. Layer growth is observed in almost all the cases. In some crystals spiral growth was also observed. In order to obtain more insight into growth by the spiral dislocation mechanism the crystals were etched with standard etchants and photomicrographs of the etched samples were recorded. These results are presented elsewhere (NIGLI et al. (c)).

References

BEST, K. J.: Phys. Kondens. Materie **1**, 316 (1963)

BROWN, F. C.: J. Phys. Chem. Solids **4**, 206 (1958)

BUROFF, A., SIMIDCHIEVA, P., STOYCHEVA, R., MALINOWSKY, J.: Model investigations of the Photographic Process, Sunchev Byrag, Bulgaria (1973)

CLARK, P. V. McD., MITCHELL, J. W.: J. Photogr. Sci. **4**, 1 (1956)

GAHWILLER, Ch., HARBECKE, G.: Phys. Rev. **185**, 1141 (1969)

GMELINS Handbuch der Anorganischen Chemie, Weinheim 1958

GOTO, T., UETA, M.: J. Phys. Soc. Japan **29**, 1512 (1970)

JONES, P. L., SITTING, C. N. W., MASON, D. E., WILLIAMS, V. A.: J. Phys. D. **1**, 283 (1968)

LYSKOVICH, A. B., SHEVKIR, Z. L., PENTSAK, G., LEREBETSKI, S. K.: Prib. i. Teckh. Eksper. **6**, 198 (1968)

NIGLI, S.: Ph. D. Thesis, University of Delhi, 1986

NIGLI, S., CHADHA, G. K., TRIGUNAYAT, G. C., BAGAI, R. K.: J. Cryst. Growth (in press) (a)

NIGLI, S., JAGADISH, C., CHADHA, G. K. (unpublished) (b)

NIGLI, S., JAGADISH, C., CHADHA, G. K. (unpublished) (c)

PANDEY, D., KRISHNA, P.: Phil. Mag. **31**, 1113 (1975) (a)

PANDEY, D., KRISHNA, P.: J. Cryst. Growth **31**, 66 (1975) (b)

PANDEY, D., KRISHNA, P.: Mater. Sci. Engg. **20**, 66 (1975) (c)

PANDEY, D., KRISHNA, P.: Mater. Sci. Engg. **26**, 53 (1976)

SHAH, J. S.: in: Crystal Growth, ed. B. R. PAMPLIN, Oxford 1980

Tubbs, M. R.: J. Phys. Chem. Solids **29**, 1191 (1968)
Tubbs, M. R.: Photogr. Sci. Eng. **18**, 156 (1974)
Wright, D. K., Tubbs, M. R.: phys. stat. sol. **37**, 551 (1970)

(Received May 14, 1987)

Authors' address:

c/o Dr. Selina Nigli
99, Priyadarshini Apartments
A-4 Paschim Vihar
New Delhi — 110063, India

Cryst. Res. Technol.	22	1987	12	K224–K226

A. J. HORODECKI

Institute of Experimental Physics
University of Gdańsk, Poland

On the Possibility of Polytypism in Semistatistically Defected Wurtzite-type Crystals of CdS

The analysis of extraordinary reflections of X-ray pattern of CdS thin films is presented which indicates the existence of 22 H polytype. For the conditions that are favourable for polytypism a new dependence of Madelung constant on the crystal structure is proposed. All of the results are consistent with the concept of semistatistically defected wurtzite crystal.

Die Analyse außergewöhnlicher Reflexe von Röntgendiagrammen von CdS-Dünnschichten zeigt die Existenz eines 22 H Polytyps an. Für die für Polytypismus günstigen Bedingungen wird eine neue Abhängigkeit der Madelung-Konstanten von der Kristallstruktur vorgeschlagen. Die gesamten Ergebnisse stimmen mit dem Konzept der semistatistisch gestörten Wurtzit-Kristalle überein.

1. Introduction

As is well known, a number of layer structures exhibits polytypism (e.g. VERMA, KRISHNA). Moreover, it does not exist any complete theory of this phenomenon. KLEBER has suggested its association with spiral growth. Recently, dependence of polytype structure on the ionicity of mixed cadmium dihalides crystals have been reported (SHARMA et al.). Nevertheless, the simplest basis for investigation of polytypism seems to be the family of AII–BVI compounds.

According to DÄWERITZ, the appearance of polytypism in these compounds is the more probable the less is difference between lattice energies of zincblende and wurtzite structure. Regarding its data due to CdS the slight decreasing of lattice energy of wurtzite is necessary for this effect. His calculations are based on the discussion of the small variations both u-coordinate and axial ratio of wurtzite that affect the value of Madelung constant. These two factors, however, can vary independly due to nonequivalent types of A–B bonds being of equal or different lenght, respectively.

Considering the semistatistically defected wurtzite-type crystal of CdS (HORODECKI (b)), its ideal lattice (ISW-hereafter) should be primarily analysed in order to estimate its Madelung constant. Then we introduce here the simple relation between Madelung constant and crystal of ideal structure of AB type. The results that are discussed together with experimental data on CdS thin films of earlier work (HORODECKI et al.) indicate the coexistence of ISW with 22 H polytype.

2. Crystal structure parameter of AB compounds

Let us viewing of crystal of AB type as consisted of a number of bonds which join the nearest neighbour cations and anions into the pairs. Notice, that besides the well known definition of number of such pairs, n, which belong to the unit cell there is another number of corresponding equivalent bonds, N which also belongs to this cell. Regarding V, the volume of the latter, it can be described as fr_0^3 i.e. related to the cube of length of bond, r_0. Thus we can define the structural parameter of crystal of AB type as follows

$$T = \frac{N}{nf}. \tag{1}$$

Very roughly, T depends on the periodical condensation of bonds in crystal space. The greater T, the less condensed bonds around the ionic sites, the greater long range interaction of ionic charges, the greater α, the Madelung constant. The data collected in Table 1 confirm the above consideration. It occurs that α is an increasing function of T. Despite of rather complicated run of the latter, the linear extrapolation of data both of ideal wurtzite and zincblende gives for ISW $\alpha = 1.6360$. Then the slight decreasing of energy of ISW lattice in comparison with that of ideal wurtzite can be expected and consequently the possibility of the polytypism occurring.

Table 1
Madelung constant, α, as dependent on the structural parameter T. The values of α are due to the cation-anion distance, r_0

Type of structure	N	n	f	T	α
CsCl	1	1	$8\sqrt{3}/9$	0.650	1.7627*)
NaCl	12	4	8	0.375	1.7476*)
ideal wurtzite	3	2	$32\sqrt{3}/9$	0.244	1.6413**)
zincblende	7	4	$64\sqrt{3}/9$	0.142	1.6381**)
ideal SW	13	7	$128\sqrt{3}/9$	0.075	1.6360***)

*) after SEITZ
**) after DÄWERITZ
***) linear extrapolation from two preceding structures

3. Extraordinary reflections of thin films of CdS

Let us assume the three extraordinary reflections (EOL- in the earlier work, HORODECKI et al.) being (00.1) reflections of the unknown polytype of CdS. Fortunately, these reflections although weak, are very narrow, so we can determine their Bragg angles due to method of asymmetric X-ray pattern (DOGIL et al.). Then we obtain $\vartheta_\alpha = 13.91°$, $\vartheta_\beta = 16.77°$ and $\vartheta_\gamma = 18.20°$ if, according to the above assumption $l = 20$, 24, and 26, respectively. Thus, the Bragg equation takes the form

$$2c_{op} \sin \vartheta = l\lambda\,, \tag{2}$$

and we obtain for c_{op} 74.468, 74.454, and 74.472 Å, respectively. By simple averaging $c_{op} = 74.465$ Å and it can be interpreted as the period of 22 H polytype if, however, the height of hexagonal packing equals 6.7695 Å. Then from equation (2) results $\vartheta_{00.2} = \vartheta_{00.22} = 15.33°$. This coincidence makes impossible to observe independly (00.22) reflection. Moreover, from the least square method used for the common wurtzite reflections $a_0 = 4.1294$ Å.

Summarizing, the films of CdS which belong to phase II (HORODECKI (a)) exhibit besides SW characteristics the polytype reflections with simultaneous increasing of axial ratio as well the volume of unit cell. The latter conclusion is due to the data of pure CdS which are $c_0 = 6.7162$ Å and $a_0 = 4.1360$ Å (CHERIN et al.). These results are in accordance with SW concept because the presence of VP is favourable for the lattice expansion. However, these effects are presumably due to the increase of c_0, so we can suppose that VP are mostly oriented along c-axis.

4. Conclusions

The significant density of VP leads not only to concept of SW (HORODECKI (b)) but also facilitates in-plane rotation i.e. supports tendency to polytypism. On the other hand, SW has lower T parameter than lower Madelung constant and, in consequence the lower lattice energy. According to DÄWERITZ the latter fact is favourable for polytypism. Nevertheless, all of these results need further investigations. Especially, the new dependence of Madelung constant on the crystal structure seems to be of interest and would be recognized in case of the real wurtzite. However, the lattice energy of ISW in terms of spatial distribution of crystal bonds will be analysed in a separate paper.

References

CHERIN, P., LIND, E. L., DAVIS, E. A.: J. Electrochem. Soc. **117**, 233 (1970)
DÄWERITZ, L.: Kristall und Technik **6**, 101 (1971)
DOGIL, B., HORODECKI, A. J., PRECHT, W.: Kristall und Technik **12**, K73 (1977)
HORODECKI, A. J., ŁEPEK, M., PRECHT, W.: Kristall und Technik **13**, K18 (1978)
HORODECKI, A. J.: Cryst. Res. Technol. (in press (a))
HORODECKI, A. J.: Cryst. Res. Technol. (in press (b))
KLEBER, W.: An Introduction to Crystallography, Berlin 1970
SEITZ, F.: The Modern Theory of Solids, London 1940
SHARMA, G. L., SHARMA, S. D., AGARWAL, V. K.: phys. stat. sol. (a) **95**, 57 (1986)
VERMA, A. R., KRISHNA, P.: Polimorfism i politypism w kristałłach, Moskwa 1969

(Received May 18, 1987)

Author's address:

Dr. A. J. HORODECKI
Institute of Experimental Physics
University of Gdańsk
ul. Wita Stwosza 57
80952 Gdańsk, Poland

Cryst. Res. Technol.	22	1987	12	K227–K229

A. J. Horodecki

Institute of Experimental Physics
University of Gdańsk, Poland

The Possible Influence of Electric Moments on the Growth of Thin Films of Pyroelectrics from Vapour Phase

Electric moments of substrate surface interact with any polar species coming-in from vapour phase. The re-mapping of substrate electricity will be continued if both these species and crystals are of uniaxial polarity. Consequently, the additional field at the growing surface is the resultant of crystal "memory" and of pyroelectric charge and seems to be the main reason for the texture. Moreover, considering the microrelaxation of growth the numerous pairs of cation vacancy — anion vacancy are expected. The results indicate the way for understanding the peculiarities of X-ray pattern of CdS thin films.

Die elektrischen Momente von Substratoberflächen beeinflussen jede polare Spezies, die sich aus der Dampfphase abscheidet. Wenn diese Spezies und auch der Kristall einachsige Polarität besitzen, so wird das Erscheinen der Substrat-Elektrizität fortgesetzt. Das zusätzliche elektrische Feld auf der wachsenden Oberfläche ist das Ergebnis des „Kristallgedächtnisses" und der pyroelektrischen Ladung und scheint die Hauptursache für die Textur zu sein. Wegen der Mikrorelaxation des Aufwachsens ist die zahlreiche Bildung von Cadmium- und Schwefellücken zu erwarten. Die Ergebnisse zeigen den Weg zum Verständnis der Besonderheiten der Röntgendiagramme von CdS-Dünnschichten.

1. Introduction

During the past decade cadmium sulphide has received permanent attention due to the experimental progress in studies of optical bistability (OB) (Reinecke; Müller et al.; Rossmann et al.; Henneberger, Rossmann; Wegener et al.). In these experiments vapour-phase-grown CdS platelets with thickness of some micrometers are used. Regarding the practical perspective OB for laser technique and optical signal processing the renewal attempts would be paid for the vapour-phase growing of CdS thin films. In particular, the consideration of vacuum deposition data only cannot explain the whole of question because the own properties of the deposited compound would be taken into account. Such the most evident property is the uniaxiality of crystal and this is case of CdS of wurtzite type.

In our earlier work the results of investigations of CdS thin films were reported (Horodecki et al.). Using the glass or mica as the substrate and ranging temperature from 373 to 473 K, two types of real structure have occurred which can be designed as phases I and II. Phase I is poorly crystallized and its X-ray pattern is powder-like. Contrary, phase II is well crystallized and quite unlike to powder due to (i) texture, (ii) strong intensities of revealed reflections and, (iii) presence of extraordinary reflections. Moreover, the medial phase M — hereafter can be distinguished which is rather similar to I except of its anomalously increased (10.3) intensity.

With regarding the above results, our aim is to indicate the possible influence of electric moments on the vapour growth of pyroelectric thin film.

2. Inclination to textured growth of film

The main features which would be analysed are as follows

(i) the occurrence of regions of substrate in which the average electric moment is different from zero (EMS — hereafter),
(ii) the appearance of the constant electric moment of the species coming-in to the substrate,
(iii) the unique polar axis of crystal.

Further, we discuss only the species being randomly oriented diatomic molecules (EMM — hereafter). Now, when they are seeding on the substrate they would be more or less ordered in accordance with the local EMS. Their movement in substrate plane as well the reevaporation cannot completely destroy this ordering because the significant rotation of EMM contrary to the field resulting both from EMS and other ordered EMM is not very probable. It follows that if at beginning the deposition the islands are formed they conserve and enhance the local EMS. Continuing deposition the group of islands with similar directions of their electric moment will behave as whole as new average electric moment of crystal (EMC — hereafter). This will be possible presumably due to the cooperation of the uniaxial polarity both of crystal and its building unit i.e. diatomic polar molecule. Indeed, in the case of the non-uniaxial polar class and multipolar building unit the thermal agitation can easier turn out the latter to another polar direction thus promote a new, accidental direction of growth.

The above described evolution of EMC depends, of course, on details of the distribution of EMS, so can result finally in textured as well in purely isotropic growth of film.

3. Production of vacancy pairs

Let us assume as previously the polar diatomic molecules coming-in with equiprobable orientation of their EMM with respect to the film surface. Some of these would be significantly reoriented according to the local EMC. But from time to time this reorientation is so delayed that a layer of growing sample will remain without a number of molecules i.e. will include the vacancy pairs (VP — hereafter). Moreover, despite the inconspicuous significance of pyroelectricity it cannot be neglected if the crystal is strongly defected, so EMC consists partly of pyroelectric moment (PM — hereafter).

In fact, the significant difference of temperature between vapour source and substrate as well the kinetic energy of seeding molecules can be cause of the gradient of temperature at the film surface. Regarding the above discussion, the significant density of VP and subsequently essential change in structure amplitude characteristics can be expected.

4. Conclusions

Phase I is then obtained if the substrate has not any sufficient EMS and crystal building unit has more than one polar direction. If, however, the latter condition is not fulfilled, phase M will be appear. In this case, although the strong disordering of

EMC excludes any texture, the production of VP is still possible as well PM can be present. The anomalous increase of (10.3) intensity which characterizes phase M would be in accordance with the above supposition. Viewing these considerations, it is worth of notice the important question of "spectrum" of building units for the investigation of real structure of vapour grown thin film. Regarding the habit of single crystal of CdS, an interesting discussion of such "spectrum" is given in the paper of Bulakh.

Summarizing, EMS would be responsible, first of all, for the texture "nucleation" of pyroelectric film as well be favourable indirectly for the high density of VP. The latter, however, can involve a "feedback" with regarding PM thus affect the initial influence of EMS. Besides, VP presence makes probable the rotation of packing i.e. occurring of polytypism. Provided the extraordinary reflections are due to a polytype, all of the peculiarities of phase II would be confirm our considerations. The answer for the latter question will be given in another report. Nevertheless, following our assumptions, the particular case of VP distribution will led us in next paper to the concept of semistatistically defected wurtzite crystal.

References

Bulakh, B. M.: J. Cryst. Growth **5**, 243 (1969)
Henneberger, F., Rossmann, H.: phys. stat. sol. (b) **121**, 685 (1984)
Horodecki, A. J., Łepek, M., Precht, W.: Kristall und Technik **13**, K18 (1978)
Müller, G. O., Weber, H. H., Horicke, I.: phys. stat. sol. (b) **91**, 531 (1979)
Reinecke, T. L.: Phys. Rev. B **16**, 2653 (1977)
Rossmann, H., Henneberger, F., Voigt, J.: phys. stat. sol. (b) **115**, K63 (1983)
Wegener, M., Klingshirn, C., Koch, S. W., Banyai, L.: Semicond. Sci. Technol. **1**, 366 (1986)

(Received May 18, 1987)

Author's address:

Dr. A. J. Horodecki
Institute of Experimental Physics
University of Gdańsk
ul. Wita Stwosza 57
80952 Gdańsk, Poland

Cryst. Res. Technol.	22	1987	12	K230–K233

LI JIANMIN, PAN GUOQIANG, FANG JIFA

Department of Modern Chemistry, China University of Science and Technology, Hefei, Anhui, P.R. China

d–d Spectra of CaM(II)-EDTA Crystals(II) Ca[Ni(EDTA)] · 4 H_2O

Introduction

Recently we have measured a series of the d–d transition spectra of CaM(II)-EDTA and given a reasonable explanation with the theory proposed by us. We have reported the electronic absorption spectrum of Ca[Co(EDTA)] · 5 H_2O crystal in the previous paper. In this paper we will continue to treat the crystal of Ca[Ni(EDTA)] · 4 H_2O in same way.

Experimental

1. The preparation of Ca[Ni(EDTA)] · 4 H_2O crystal

The crystal was prepared by mixing aqueous solution of EDTA, calcium chloride and nickel chloride with a ratio 1:1:1. The solution was slowly evaporated at 60 °C, then was placed 2–3 days at room temperature. A deep green crystal was obtained. The size of the crystal was about 0.5–1.5 mm. It is confirmed by chemical element analysis that the crystal is Ca[Ni(EDTA)] · 4 H_2O.

2. The measurement of the d–d spectrum

The electronic absorption spectrum of Ca[Ni(EDTA)] · 4 H_2O crystal was measured in the region of 4000–30000 cm^{-1}. The result measured is shown in Figure 1. There are strong absorption peaks at 9000, 13000, 16000, 25000 cm^{-1}, respectively.

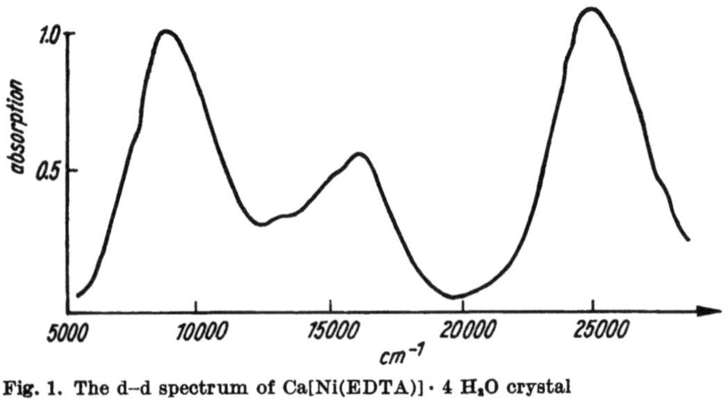

Fig. 1. The d–d spectrum of Ca[Ni(EDTA)] · 4 H_2O crystal

Theoretical explanation

According to the point charge-point dipole model within the framework of ligand theory we calculated the d–d transition energy levels of Ca[Ni(EDTA)] · 4 H_2O. From the data of the crystal structure (NESTEROVA 1984) it can be seen that the symmetry of Ca[Ni(EDTA)] · 4 H_2O crystal is very low. We chose O(3)–Ni–O(7) as z-axis, the vertical bisector of N(1)–Ni–N(2) as x-axis, then the y-axis is determined by the dextral rule (see Fig. 2). The symmetry of Ni(II) and its first neighbour ligands belong

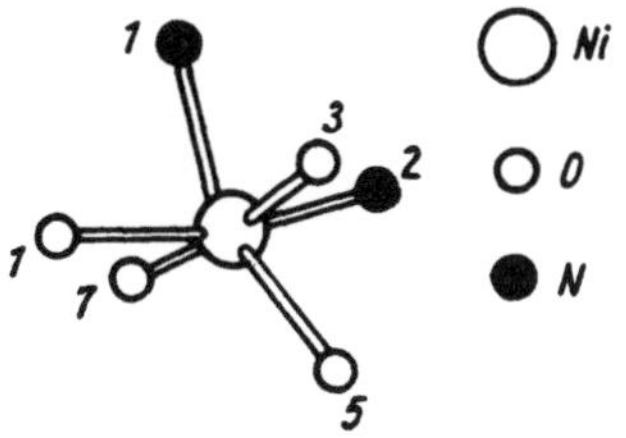

Fig. 2. The crystal structure of Ca[Ni(EDTA)] · 4 H_2O

to C_2 group. The electronic configuration of the central Ni(II) ion is d^8. In the ligand field its ground state ^3F is split into three ^3A and four ^3B, ^3P similarly into one ^3A and two ^3B. So the orbital degeneracy breaks down in this low symmetric ligand field. We calculated the matrix elements of the ligand field operator as follows: (weak field scheme): (see page K232)

where

$$\begin{cases} a_1 = \dfrac{2}{7}\left[\dfrac{(3\cos^2\theta_1 - 1)}{\bar{R}_1^3} + \dfrac{(3\cos^2\theta_2 - 1)}{\bar{R}_2^3} + \dfrac{2}{\bar{R}_3^3}\right] K\bar{N}^2 < \langle r^2\rangle\, P^{(2)} \\ a_2 = \dfrac{2}{7} K\bar{N}^2\, P^{(2)}\langle r_2\rangle \left(\dfrac{\sin\theta_1 \cos\Phi_1}{\bar{R}_1^3} + \dfrac{\sin^2\theta_2 \cos\Phi_2}{\bar{R}_2^3}\right) \\ a_3 = \dfrac{2}{21} K\bar{N}^2\, P^{(4)}\langle r^4\rangle \left(\dfrac{35\cos^4\theta_1 - 30\cos^2\theta_1 + 3}{\bar{R}_1^5} + \dfrac{35\cos^4\theta_2 - 30\cos^2\theta_2 + 3}{\bar{R}_2^5} + \dfrac{8}{\bar{R}_3^5}\right) \\ a_4 = \dfrac{2}{21} K\bar{N}^2\, P^{(4)}\langle r^4\rangle \left[\dfrac{\sin^2\theta_1(7\cos^2\theta_1 - 1)\cos\Phi_1}{\bar{R}_1^5} + \dfrac{\sin^2\theta_2(7\cos^2\theta_2 - 1)\cos\Phi_2}{\bar{R}_2^5}\right] \\ a_5 = \dfrac{2}{21} K\bar{N}^2\, P^{(4)}\langle r^4\rangle \left[\dfrac{\sin^4\theta_1 \cos 2\Phi_1}{\bar{R}_1^5} + \dfrac{\sin^4\theta_2 \cos^2\Phi_2}{\bar{R}_2^5}\right]. \end{cases} \quad (2)$$

In terms of the scaling wave function theory of non-free ion, the double-ζ radial wave function of Ni(II) can be written as: (Zang RONGCHUN, Li JIANMIN 1982)

$$R(r, \Omega) = c^{-1/2}\left\{0.534855\left[\frac{(2\zeta_1)^7}{6!}\right]^{1/2} r^2 e^{-\zeta_1 r} + 0.62500\left[\frac{(2\zeta_2)^7}{6!}\right]^{1/2} r^2 e^{-\zeta_2 r}\right\} \quad (3)$$

where

$$\begin{cases} \zeta_1 = 6.1282(1 - 0.351045\Omega - 0.208760\Omega^2)\,, \\ \zeta_2 = 2.4250(1 - 1.584119\Omega + 0.322824\Omega^2)\,, \\ C = 1 - 0.668569\left\{0.483577 - \left[\dfrac{2\sqrt{(\zeta_1\zeta_2)}}{(\zeta_1 + \zeta_2)}\right]^7\right\}. \end{cases} \quad (4)$$

$$
\begin{array}{cccc}
|^3A(^3T_{2g},{}^3F)\rangle, & |^3A(^3T_{2g},{}^3F)\rangle, & |^3A(^3T_{1g},{}^3F)\rangle, & |^3A(^3T_{1g},{}^3P)\rangle
\end{array}
$$

$$
\begin{bmatrix}
-\frac{7}{8}(a_3+5a_5) & 0 & 0 & 0 \\
 & \frac{3}{4}a_2+\frac{7}{32}a_3+\frac{5}{4}a_4+\frac{105}{32}a_5, & \sqrt{\frac{3}{20}}\left(a_1-\frac{a_2}{2}+\frac{5a_3}{16}+5a_4+\frac{35}{16}a_5\right), & \sqrt{\frac{3}{5}}\left(a_1+2a_2+\frac{5}{16}a_3+\frac{15}{8}a_4+\frac{35}{16}a_5\right) \\
 & & \frac{1}{5}a_1-\frac{3}{20}a_2+\frac{9}{32}a_3-\frac{15}{4}a_4-\frac{105}{32}a_5, & \frac{3}{5}a_1-\frac{6}{5}a_2+\frac{7}{16}a_4-\frac{35}{16}a_5 \\
 & & & -\frac{7}{10}a_1-\frac{21}{10}a_2+15\,\overline{N}^4B
\end{bmatrix}
$$

$$
\begin{array}{l}
|^3B(^3A_{1g},{}^3F)\rangle \\
|^3B(^3T_{1g},{}^3P)\rangle \\
|^3B(^3T_{2g},{}^3F)\rangle
\end{array}
\begin{bmatrix}
-\frac{1}{8}(a_3-5a_5) & -\sqrt{\frac{3}{5}}\left(a_2-\frac{5}{4}a_4\right) & 2\sqrt{\frac{3}{5}}\left(a_2-\frac{5}{4}a_4\right) \\
 & -\frac{2}{5}a_1+\frac{3}{4}a_3 & -\frac{6}{5}\left(a_1+\frac{1}{2}a_3\right) \\
 & & \frac{1}{5}a_1+15\overline{N}^4B
\end{bmatrix}
$$

$$
\begin{array}{ccc}
|^3B(^3T_{2g},{}^3F)\rangle & |^3B(^3T_{1g},{}^3F)\rangle & |^3B(^3T_{1g},3P)\rangle
\end{array}
$$

$$
\begin{bmatrix}
-\frac{3}{4}a_2+\frac{7}{32}a_3-\frac{5}{4}a_4+\frac{105}{32}a_5, & -\sqrt{\frac{3}{20}}\left(a_1+\frac{1}{2}a_2+\frac{5}{16}a_3-5a_4+\frac{35}{16}a_5\right), & \sqrt{\frac{3}{5}}\left(a_1-2a_2+\frac{15}{16}a_3-\frac{15}{8}a_4+\frac{35}{16}a_5\right) \\
 & \frac{a_1}{5}+\frac{3}{20}a_2+\frac{9}{32}a_3+\frac{15}{4}a_4-\frac{105}{32}a_5, & \frac{3}{5}a_1+\frac{6}{5}a_2+\frac{3}{16}a_3-\frac{5}{8}a_4+\frac{35}{16}a_5 \\
 & & -\frac{1}{10}a_1+\frac{21}{10}a_2+15\overline{N}^4B,
\end{bmatrix}
\quad (1)
$$

From the formula (3), the radial wave function of Ni(II) in various ligand environment can be obtained, for the $\langle r^2 \rangle$, $\langle r^4 \rangle$, Racah parameter B and C can be calculated by it. The coefficients a_1, a_2, a_3, a_4, a_5 Racah parameter B and C are calculated with the average environment rule (ROSENBERG et al.) because Ni(II) ion is mixedly coordinated by oxygen and nitrogen. Using the data above we calculated the d–d transition energy

Table 1
The d–d spectrum of Ca[Ni(EDTA)] · 4 H_2O crystal

assignment	calc.	obs.	assignment	calc.	obs.
$^3B(^3A_{2g}, {}^3D)$	0		$^3B(^3T_{1g}z, {}^3F)$	14830	15200
$^3A(^3T_{2g}\xi, {}^3F)$	8320	7500	$^3B(^3T_{1g}x, {}^3F)$	15600	16150
$^3A(^3T_{2g}\zeta, {}^3F)$	8950	8850	$^3B(^3T_{1g}x, {}^3P)$	23230	23200
$^3B(^3T_{2g}\eta, {}^3F)$	9470	8850	$^3B(^3T_{1g}z, {}^3P)$	25740	24950
$^3A(^3T_{1g}y, {}^3F)$	13830	13300	$^3A(^3T_{1g}y, {}^3P)$	28520	27850

levels of the triplet of Ni(II). The results calculated are listed in Table 1. They are in good agreement with the experimental values. This method both retains the conciseness of the point charge model in the ligand field theory and partially considers the non-local characterastic in MO theory. It is quite practical for explanation of the d–d spectrum and ESR even Mössbauer spectrum of the coordination compound (ZHANG YONGFENG, Li JIANMIN 1985 et al.).

References

NESTEROVA, Ya. M.: Koord. Khimiya **10** (1), 129 (1984)
ROSENBERG, R. C., ROOT, C. A., GRONY, H. B.: J. Amer. Chem. Soc. **97**, 21 (1975)
ZANG RONGCHUN, LI JIANMIN, J.: China Univ. Sci. Tech. **12**, 74 (1982)
ZHANG YONGFENG, LI JIANMIN, HUANG DEBIN, HUANG JILING: Scientia Sinica (Series B) **28** (12), 1257 (1985)

(Received May 21, 1987)

Authors' address:

Ass. Professor LI JIANMIN, PAN GUOQIANG, FANG JIFA
University of Science and Technology of China
Department of Modern Chemistry
Hefei, Anhui, 230029
P.R. China

Cryst. Res. Technol.	22	1987	12	K234–K236

LI JIANMIN, PAN GUOQIANG, FANG JIFA

Department of Modern Chemistry, China University of Science and Technology, Hefei, Anhui, P. R. China

d–d Spectra of CaM(II)-EDTA Crystals (III) Ca[Cu(EDTA)] · 4 H_2O

Introduction

The crystal of Ca[Cu(EDTA)] · 4 H_2O just like the crystals of Ca[Co(EDTA)] · 5 H_2O and Ca[Ni(EDTA)] · 4 H_2O we reported, its symmetry is also very low, but we still make a reasonable explanation for its d–d spectrum with the theory developed by us.

Experimental and results

The crystal was prepared by mixing aqueous solution of EDTA, calcium chloride and copper chloride with a molar ratio 1:1:1. The solution was slowly evaporated at 60 °C, then was placed several days at room temperature. A blue crystal was obtained. The size of the crystal is about 1 – 2 mm. It is confirmed by chemical element analysis that the crystal is Ca[Cu(EDTA)] · 4 H_2O.

The electronic absorption spectrum of Ca[Cu(EDTA)] · 4 H_2O crystal was measured in the region of 5000 – 20000 cm^{-1}. The result measured is shown in Figure 1.

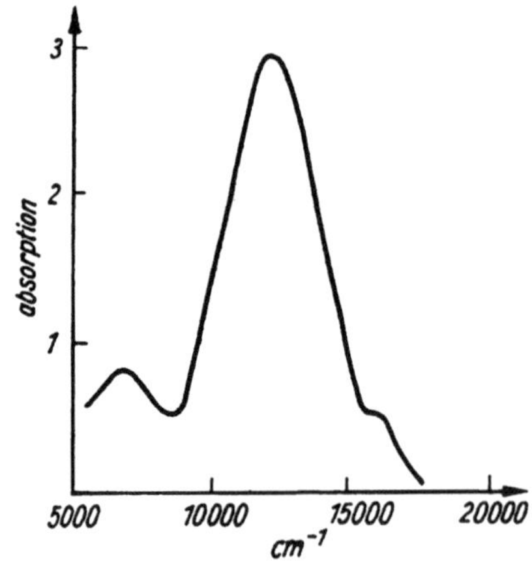

Fig. 1. The d–d spectrum of Ca[Cu(EDTA)] · 4 H_2O crystal

Theoretical explanation

From the crystal structure determined by NESTEROVA et al. It can be seen that the symmetry of Ca[Cu(EDTA)] · 4 H_2O crystal is very low (see Fig. 2). The first neighbour coordination atoms of Cu(II) are two nitrogen atoms and four oxygen atoms, but the bond length of Cu(II)–O(5) is specially long (2.711 Å). The effect of O(5) is much weaker

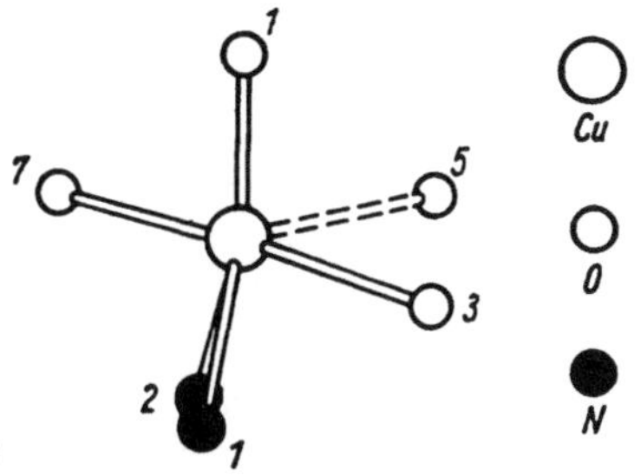

Fig. 2. The crystal structure of Ca[Cu(EDTA)] · 4 H_2O

than other atoms. So we can ignore it. Thus the approximate symmetry of Cu(II) and its first neighbour ligands belong to Cs group. The energy matrix of d^9 configuration in Cs crystal field (weak field scheme) is:

$$^2A'(^2T_{2g}, z^2D) = a_1 - \frac{1}{8}a_3 + \frac{35}{8}a_5 ,$$

$$^2A'(^2T_{2g}, y, ^2D) = -\frac{1}{2}a_1 - \frac{3}{2}a_2 + \frac{1}{2}a_3 - \frac{5}{2}a_4 ,$$

$$^2A'(^2T_{2g}, x, ^2D) = -\frac{1}{2}a_1 + \frac{3}{2}a_2 + \frac{1}{2}a_3 + \frac{5}{2}a_4 ,$$

$$\begin{array}{l} ^2A'(^2E_g\ \theta.\ ^2D) \\ ^2A'(^2E_g\ \varepsilon,\ ^2D) \end{array} \begin{bmatrix} -a_1 - \frac{3}{4}a_3 & \sqrt{3}\left(a_2 - \frac{5}{4}a_4\right) \\ & a_1 - \frac{1}{8}a_3 - \frac{35}{8}a_5 \end{bmatrix}, \tag{1}$$

where

$$a_1 = \frac{1}{7} K\bar{N}^2 P^{(2)}\langle r^2\rangle \left(\frac{4}{R_{11}^3} - \frac{1}{R_1^3} - \frac{1}{R_2^3} - \frac{1}{R_3^3}\right),$$

$$a_2 = \frac{1}{7} K\bar{N}^2 P^{(2)}\langle r^2\rangle \left(\frac{1}{R_1^3} + \frac{\cos^2 2\Phi_1}{R_2^3} + \frac{\cos 2\Phi_2}{R_3^3}\right),$$

$$a_3 = \frac{1}{21} K\bar{N}^2 P^{(4)}\langle r^4\rangle \left[\frac{16}{R_{11}^5} + 3\left(\frac{1}{R_1^5} + \frac{1}{R_5^2} + \frac{1}{R_3^5}\right)\right],$$

$$a_4 = -\frac{1}{21} K\bar{N}^2 P^{(4)}\langle r^4\rangle \left(\frac{1}{R_1^5} + \frac{\cos 2\Phi_1}{R_2^5} + \frac{\cos 2\Phi_2}{R_3^5}\right)$$

$$a_5 = \frac{1}{21} K\bar{N}^2 P^{(4)}\langle r^4\rangle \left(\frac{1}{R_1^5} + \frac{\cos 4\Phi_1}{R_2^5} + \frac{\cos 4\Phi_2}{R_3^5}\right). \tag{2}$$

According to the scaling wave function theory of non-free ion, the double-ζ radial wave function of Cu(II) can be written as: (ZHANG YONGFENG, LI JIANMIN 1983)

$$R(r, \Omega) = C^{-y_2}\left\{0.55428 \sqrt{\frac{2(\zeta_1)^7}{6!}}\, r^2 e^{-\zeta_1 r} + 0.60500 \sqrt{\frac{2(\zeta_2)^7}{6!}}\, r^2 e^{-\zeta_2 r}\right\}, \tag{3}$$

where

$$\zeta_1 = 6.3496(1 - 0.4150\Omega - 0.2500\Omega^2) ,$$

$$\zeta_2 = 2.5250(1 - 1.6450\Omega + 0.3550\Omega^2) ,$$

$$C = 1 - 0.67078\left[0.48719 - \left(\frac{2\sqrt{\zeta_1\zeta_2}}{\zeta_1 + \zeta_2}\right)^7\right]. \tag{4}$$

Using the formula (3), the radial wave function of Cu(II) ion in various ligand environment can be obtained, for the $\langle r^2\rangle$, $\langle r^4\rangle$, Racah parameter B and C can be calculated by it. The coefficients a_1, a_2, a_3, a_4, a_5 and B, C are calculated with the average environment rule because Cu(II) is mixedly coordinated by oxygen and nitrogen. From the data of the crystal structure we obtain $R_1 = 1.9686$ Å, $R_2 = 2.3445$ Å, $R_3 = 2.0765$ Å, $R = 1.990$ Å, $\Phi_1 = 101.4°$, $\Phi_2 = 174.58°$. We calculated the energy levels of $Ca[Cu(EDTA)] \cdot 4\,H_2O$ crystal field, the results calculated are listed in Table 1. They are good in agreement with the experimental results. It is difficult to give good theoretical explanation with a general method because this compound not only is mixedly coordinated but has very low symmetry.

Table 1
The d–d spectrum of $Ca[Cu(EDTA)] \cdot 4\,H_2O$ crystal

assignment	calc.	obs.	assignment	calc.	obs.
$^2A'(^2E_g\theta, {}^2D)$	0		$^2A'(^2T_{2g}z, {}^2D)$	15850	16100
$^2A'(^2E_g\varepsilon, {}^2D)$	7530	7000	$^2A'(^2T_{2g}x, {}^2D)$	20870	
$^2A'(^2T_{2g}y, {}^2D)$	12230	12250			

References

Nesterova, Ya. M.: Koord. Khimiya **10** (1), 129 (1984)
Rosenberg, R. C., Root, C. A., Grony, H. B.: J. Amer. Chem. Soc. **97**, 21 (1975)
Zhang Yongfeng, Li Jianmin: J. Struct. Chem. (China) **2** (3), 189 (1983)

(Received June 2, 1987)

Authors' address:

Ass. Professor Li Jianmin, Pan Guoqiang, Fang Jifa
University of Science and Technology of China
Department of Modern Chemistry
Hefei, Anhui, 230029
P.R. China

Cryst. Res. Technol.	22	1987	12	K237

V. Pučálka

University of Mining and Metallurgy, Ostrava, ČSSR

Another Equation for Interplanar Spacings

The vector equation of a lattice plane (*hkl*) lying in a real space may be derived from the condition that the volume of a parallelepiped constructed from vectors lying in this plane equals zero. By means of vectors, a lattice plane is given by three position vectors $\boldsymbol{a}/h$, $\boldsymbol{b}/k$, $\boldsymbol{c}/l$ and the position vector $\boldsymbol{r}$ of an arbitrary point lying in this plane. For the volume the following holds:

$$\left(\frac{\boldsymbol{a}}{h} - \boldsymbol{r}\right)\left[\left(\frac{\boldsymbol{b}}{k} - \boldsymbol{r}\right) \times \left(\frac{\boldsymbol{c}}{l} - \boldsymbol{r}\right)\right] = 0 .$$

Having done the indicated products and some arrangements, we obtain the following equation of lattice plane

$$[h(\boldsymbol{b} \times \boldsymbol{c}) + k(\boldsymbol{c} \times \boldsymbol{a}) + l(\boldsymbol{a} \times \boldsymbol{b})] \cdot \boldsymbol{r} - \boldsymbol{a} \cdot (\boldsymbol{b} \times \boldsymbol{c}) = 0 .$$

It is an equation of type $\boldsymbol{n} \cdot \boldsymbol{r} - V = 0$ where

$$\boldsymbol{n} = h(\boldsymbol{b} \times \boldsymbol{c}) + k(\boldsymbol{c} \times \boldsymbol{a}) + l(\boldsymbol{a} \times \boldsymbol{b}) ,$$

is the equation of a normal of the lattice plane (*hkl*), and where $V = \boldsymbol{a} \cdot (\boldsymbol{b} \times \boldsymbol{c})$ is the volume of a unit cell.

As known from analytical geometry, the absolute value of the distance between the lattice plane and the origin of coordinates is $d = V/n$. The interplanar spacing is parallel to the vector of normal, so that from the *d*-value expressed by means of vectors it holds:

$$\boldsymbol{d} = \frac{V}{n^2} \boldsymbol{n} .$$

The vector of body diagonal is $\boldsymbol{D} = \boldsymbol{a} + \boldsymbol{b} + \boldsymbol{c}$, its length is

$$D^2 = a^2 + b^2 + c^2 + 2bc \cos\alpha + 2ac \cos\beta + 2ab \cos\gamma.$$

By scalar multiplication of the vector $\boldsymbol{d}$ and vector $\boldsymbol{D}$ we obtain

$$\boldsymbol{d} \cdot \boldsymbol{D} = d \cdot D \cdot \cos w ,$$

$$\boldsymbol{d}(\boldsymbol{a} + \boldsymbol{b} + \boldsymbol{c}) = d^2(h + k + l) .$$

The resulting equation is

$$d = \frac{D \cos w}{h + k + l} .$$

In this equation the geometrical parameters of the unit cell and the Miller indices are separated from each other.

(Received June 2, 1987)

Author's address:

Dr. V. Pučálka
University of Mining and Metallurgy, Department of Geology and Mineralogy
CS — 708 33 Ostrava
Czechoslovakia

Cryst. Res. Technol.	22	1987	12	K238–K240

R. N. Bekimbetov, A. A. Vaipolin, N. N. Konstantinova, L. V. Kradinova, G. A. Medvedkin, V. D. Prochukhan, Yu. V. Rud, M. A. Tairov

A.F. Ioffe Physico-Technical Institute, Academy of Sciences of the USSR, Leningrad

The Semiconductors of $Mn_2^{III}Te_4$ Family: Physical Properties, Promising Applications

Among the numerous magnetic semiconductors there is a family of ternary compounds $Mn_2^{III}Te_4$. They consist of the elements with unfilled d- and f-shells. Nowadays in scientific literature only a few works are devoted to growth of such crystals and to the physico-chemical, crystal-structural and several other properties of ternary manganese compounds, namely $InIn_2Te_4$, $MnGa_2Te_4$ (Range et al.; Panzer et al.; Babaeva et al.). For the first time the optical and electrical parameters of these materials as well as the methods of preparing of the diode structures were presented in works (Averkieva et al.; Bekimbetov et al.). Investigations of the optical, electronic, photoelectric and luminescence properties of the ternary compounds $Mn_2^{III}Te_4$ with the magnetic

Table 1

Compounds characteristics	$MnIn_2Te_4$	$MnGa_2Te_4$
melting temperature, °C	755 (Babaeva)	845 (Rustamov)
lattice parameters, Å	$a = 6.196 \pm 0.005$ $c = 12.402 \pm 0.010$	
space group	$I\bar{4}2m$	
fundamental transition	direct	indirect
values energy gap, eV E_G^{opt} (300 K) $E_G^{photoel}$ (300 K) E_G^{el} (0 K)	 1.42 1.42 1.47	 1.34 1.53
$(dE_G/dT) \cdot 10^4$ (80–300 K), eV/K	–5.3	–4.8
electrical resistivity (300 K), Ohm · cm single crystals layers	 10^2–10^6 $(2-4) \cdot 10^2$	 10^3–10^5
type of conductivity	n, p	p
carrier concentration (300 K), cm^{-3}	10^{12}–10^{13}	10^{13}
Hall's mobility, $cm^2/V \cdot s$	1–5	1
activation energy, eV	0.2–0.5	0.2–0.5
energy of photoluminescence peak (77 K), eV	1.19	
transmission range, μm	1–22	2–15
minimum absorption coefficient, cm^{-1}	0.1–1	1–2
extinction index	$(1-4) \cdot 10^{-3}$	
reflectance coefficient, %	25–35	25–30
refringence index	2.77–2.89 (Ge, Cu) 2.63–2.74 (Cr)	

ion Mn^{2+} are carried out in the present paper and are summarized in Table 1. Samples are obtained and characterized as to prepare single crystals and thin films.

Single crystals were grown by the method of directional crystallization from the melts of stoichiometric composition and also by containing the excess of intrinsic components of compounds. A possibility of impurity doping with elements of Cu, Ge, Cr, V and P is established. Thin films are prepared by the vacuum thermal evaporation of homophase materials on Si or SiO_2 substrates. The ways of obtaining the single crystals with n- and p-type conductivity and polycrystalline films of n-$MnIn_2Te_4$ and p-$MnGa_2Te_4$ were established. Both of ternary compounds have typically semiconducting mechanism of electrical conductivity.

It is possible to create the photosensitive structures on these ternary semiconductors of $MnIn_2Te_4$ and $MnGa_2Te_4$ such as heterojunctions on the silicon substrates, Schottky barriers with copper rectification contact and photoelectrochemical cells (Table 2).

Table 2

Diode structure parameters (300 K)	$MnIn_2Te_4$	$MnGa_2Te_4$
rectification ratio	10^4	10^2
breakdown voltage, V	~ 15	
reverse current, A ($U = 1$ V)	10^{-9}	
energy of potential barriers, eV		
semiconductors-electrolyte	0.3	0.1
Cu-semiconductors	0.4 – 0.5	0.2 – 0.3
photosensitivity, V/W	$5 \cdot 10^5$	10^2
mA/W	~ 1	~ 0.1
maximum of photosensitivity spectrum, eV	1.6	2.6
photovoltage, V	0.4	
photocurrent short-circuit, $\mu A/cm^2$ ($L = 2$ mW/cm²)	0.5	

The Schottky diode and anisotype heterojunctions Si/$MnIn_2Te_4$ are characterized by absence of the photosensitivity drop in the range from 1.45 eV to 3.0 eV. The high voltaic sensitivity and the absence of degradation of Mn(In, Ga)$_2$Te$_4$ diode structures make these materials promising for energy conversion devices, i.e. solar cells. The high optical transmission of these magnetic semiconductors may be used in various optical devices.

References

AVERKIEVA, G. K., BEKIMBETOV, R. N., KONSTANTINOVA, N. N., MEDVEDKIN, G. A., PROCHUKHAN, V. D.: Theses of Soviet Conference of Materials Science of Chalcogenide and Oxide Semiconductors, Chernovtsi. 1, 64 (1986)

BABAEVA, P. K., RUSTAMOV, P. G.: Azerb. Chem. J. 2, 124 (1983)

BABAEVA, B. K., RUSTAMOV, P. G., ALIDZHANOV, M. A.: Theses of Soviet Conference of Ternary Semiconductors and their Application, Kishinev 147, (1976)

Bekimbetov, R. N., Rud, Yu. V., Tairov, M. A.: Theses of Soviet Conference of Materials Science of Chalcogenide and Oxide Semiconductors, Chernovtsi. **1**, 104 (1986)
Panzer, B., Range, K.-J.: Acta Cryst. **C41**, 1007 (1985)
Range, K.-J., Hübner, H.-J.: Z. Naturforsch. **30b**, 145 (1975)
Rustamov, P. G., Babaeva, B. K., Adzharova, D. S.: Azerb. Chem. J. No 5, 112 (1978)

Authors' address:

R. N. Bekimbetov, Dr. A. A. Vaipolin, Dr. N. N. Konstantinova, Dr. L. V. Kradinova, Dr. G. A. Medvedkin, Prof. V. D. Prochukan, Dr. Yu. V. Rud, Dr. M. A. Tairov
Physico-Technical Institute, Academy of Sciences of the USSR
Polytechnical str. 26
194021 Leningrad
USSR

Cryst. Res. Technol.	22	1987	12	R1—R13

Crystal Research and Technology
Vol. 22 (1987)

Subject Index

Author Index

Continued from cover page 4

Short Notes

Book Reviews

Crystal Research and Technology
Volume 22 **1987** **Number 12**

Contents

Original Papers

Continued on cover page 3

Crystal Research and Technology is indexed in Current Contents/PC & ES.

Bei Fragen zur Produktsicherheit wenden Sie sich bitte an:
If you have any questions regarding product safety,
please contact:

Walter de Gruyter GmbH
Genthiner Straße 13
10785 Berlin
productsafety@degruyterbrill.com